THE
SCIENCE
GARDEN
BIODIVERSITY

THE

Julian Doberski

SCIENCE OF

GARDEN

The Living Garden

BIODIVERSITY

Pimpernel
Press Ltd
www.pimpernelpress.com

www.pimpernelpress.com

The Science of Garden Biodiversity: The Living Garden

A catalogue record for this book is available from the British Library

Cover design by Sarah Pyke
Illustrations by Thomas Bohm, User Design, Illustration and Typesetting
Typeset in Baskerville
Typesetting by Danny Lyle

ISBN 978-1-914902-09-3

Printed and bound in Great Britain
9 8 7 6 5 4 3 2 1

Contents

Chapter 1
Wildlife in a Garden

You step out of the back door into the garden.[1] Depending on the time of year, there may be signs of activity. A few birds flitting around, perhaps butterflies or bumblebees on the wing. But often the garden can seem very still. The plants are growing quietly – to do so they sense and respond to the world around them in an undemonstrative way. Some scientists would describe them as 'sentient', but that is a debate for another time and place. On the surface the garden is typically calm and still – and often thought of by gardeners and owners as a place of retreat. But, of course, if you peel back the superficial layer of inactivity, there is a web of wildlife and their ecology to explore, to understand and then to manage for a more biodiverse garden.

There are a large number of 'how to' books if you are looking for information on practical steps to maximize the wildlife potential of your garden. One of those published recently (Thomas, 2017) began by listing six myths about gardens and wildlife. I have listed just three of them here because they are particularly pertinent to this book – although my book ranges more widely into both the biodiversity and ecology of gardens.

---

1 In North America and Australasia the term 'yard' may be used in place of 'garden'. For convenience, we will use the term 'garden' throughout the book.

1   Wildlife gardens are great for wildlife – other gardens aren't.
2   A garden fit for wildlife must be 'wild'.
3   You must grow native plants.

The purpose of this book is to show how data and science can help to dispel these kinds of myths and provide an understanding of how diverse life in gardens can be and how gardens function ecologically. And along the way there will be scientific 'signposts' to better wildlife gardening.

Simply through casual observation, many gardeners will be aware of the birds, garden plants and some of the mammals found in the average garden. In addition, many people with gardens provide extra encouragement for birds to visit, with bird feeders and bird tables. These can draw down droves of birds to feed – ranging in size in the UK from blue tits and sparrows to much larger rooks, magpies and pigeons. The chances are that grey squirrels will also take part in any melee at the feeders, using a deal of acrobatic ingenuity to defeat the design of feeders meant for birds. In parts of London and increasingly spreading beyond, the sight of more exotic-looking ring-necked parakeets (originally from Asia) at the bird feeder is no longer unusual. Greener suburbs and often those not-so-green will harbour populations of foxes, badgers and Muntjac deer – the latter 'gatecrashing' from Asia. In other parts of the country, putting out meat-based food in gardens can even bring in carrion-eating red kites, spectacular birds of prey successfully reintroduced to the UK after becoming virtually extinct. As for the plants, that depends on the level of enthusiasm and preferences of the garden owner. The slightly neglected garden may be a riot of 'weedy' plants – some exotic but many native – eking out a short life before being strimmed or ending up on the compost heap. By contrast, the manicured garden may have a plethora of exotic and native plants. In both cases, some of the species involved are likely to have been subject to selective breeding – to produce varieties even more showy – and often looking flamboyantly different from the original species. In addition, gardens will typically display not only herbaceous plants but also shrubs, trees and the ubiquitous lawn.

Most gardens will have had some attention paid to design – with 'features' ranging from the simple to the grandiose. These may vary from conventional herbaceous borders to vegetable beds, rockeries, gravel beds, ponds, garden walls, statues and so on. These provide opportunities for plants with particular requirements to thrive and thrill. Even a stone statue can provide home for lichens, which otherwise may have no suitable substrate.

That, in a nutshell, perhaps summarizes what the average gardener 'sees' in a garden – from its floral variety and aesthetics to its animal wildlife. But, of course, a garden is a complex living entity. That complexity is hard to appreciate – and even documenting lists of species misses the point that a garden is more than just an inventory of what is there. One could summarize this view by noting that a garden, in functional terms, is much more than a sum of its parts. OK – that's fine as a general statement, but what does this actually mean?

Recently I wrote a short book about compost (Doberski, 2022) – with specific reference to garden compost heaps. In that book I delved into compost from an ecological perspective, noting that the compost heap is a true ecosystem. Broadly speaking, one can define an ecosystem as a largely self-contained ecological unit. Because ecosystems have no defined size or scale, we can take a few steps back from the compost heap and, broadening the scale of what we see, think of the whole garden as an ecosystem. In this case the size will be defined by the hedges or fences that define ownership. That doesn't mean that a garden will be self-contained – clearly, birds and mammals are wanderers and will move in and out of the garden. But bearing in mind the individual input of the gardener, managing their garden in a particular way, there will be a degree of ecological isolation of one garden from the next.

Once we take on board the concept of a garden ecosystem, we can start to delve into its complexity. This is not to befuddle but to try to understand the broad ecological functioning of a garden – to see it as a living 'entity' that is constantly doing things. We can draw some kind of analogy with a sleeping human body. Superficially still and inactive but with quiet breathing, a faint reminder of life – along with unseen processes such as digestion (gurgling), brain function (dreaming) and muscle action (twitching). Least evident is the overarching process of biochemical metabolism, which drives all these disparate bodily functions. This is the sum of all the chemical processes that sustain life. In a similar way, basic metabolic processes are driving life in the garden and are at the heart of ecosystem function. But we are going to start at the other end of the size spectrum – the ecological functioning of the complex of species in the garden – albeit dipping into molecules and metabolism when it helps to illuminate ecological processes. So, this is not a book about chemistry; it is a book about garden ecology. For completeness, we can remind ourselves that 'ecology' is the scientific study of the interactions between living organisms (within and between species) and with their non-living environment.

Let's imagine ourselves faced with a garden – head scratching – and thinking: how does this all work? How is it that the garden keeps going from year to year? How is the fabric of life maintained in the garden? If we spin a child's spinning top, it soon runs out of energy and stops. Yet a garden keeps going – even with minimal intervention. Light energy is constantly captured by plants and channelled through an immensely complex network of pathways through the garden ecosystem. All those ecological interactions and metabolic processes are the essence of what keeps the garden 'machine' going – albeit rather quietly and without fuss.

As we will see in later chapters, because of the large number of species found in gardens – probably many more than you imagine – we can think of the garden as our own 'nature reserve'. This has a 'natural' dynamic of its own but one on which we impose our own creativity through organizing and manipulating nature. So, as gardeners we regularly alter the path that nature would take in the garden, but still need nature to 'work' with us.

But why do we need to know how the garden 'works' eco-logically? Understanding is always a good thing per se; but beyond that, many gardeners might want to do their bit for biodiversity. One example of where gardeners can make a differ-ence relates to bee populations – which have been assailed by a range of issues, including a barrage of agrochemicals in areas of arable agriculture (Siviter et al., 2021). Knowing more of the research evidence showing how to promote garden biodiversity helps us to garden 'ecologically' without relying on the myths referred to earlier. If gardeners start to follow advice to reduce or eliminate the use of biocidal chemicals, we could increas-ingly move to the rather ironic situation where man-made city gardens become important refuge areas for a variety of species (including bees) suffering from various forms of environmental degradation beyond the boundaries of city gardens and parks. But this requires an understanding of what is important in the nurturing of garden wildlife.

It has taken quite a time, but in recent years professional ecology and ecologists have started to take a growing interest in gardens and urban environments in general. I have on my bookshelf a slightly dusty copy of the *Second European Ecological Symposium* on urban ecology, which took place as long ago as 1980 in Berlin (Bornkamm et al., 1982). This book laid down a marker for the era when urban ecology became a subject for serious academic study. Since then, a variety of publications covering urban ecology and garden wildlife have appeared, including a range of books and several scientific journals which are referenced in this book.

One of the ecological concepts that has generated a lot of relatively recent scientific interest, including in the context of gardens, is that of 'ecosystem services'. Without getting too involved in complex definitions at this stage, this concept is all about what nature does for us – often unseen and unappreciated until things go wrong. Gardeners also need ecosystem services from nature, whether it is rotting down dead vegetation in soil or predators that helpfully consume 'pest' species or bees to pollinate their bean plants. And gardens can contribute to ecosystem services beyond the boundaries of the garden. This is something discussed in more detail in Chapter 7.

Gardeners tend to think of gardens as very much their piece of green space, with thoughts seldom straying over the garden fence. But a garden is part of a complex urban jigsaw that grades from areas of high-density housing and/or commercial spaces to the lower-density suburbs, the city fringes and more open countryside. Although there is imperfect connectivity between blocks of green space (such as blocks of gardens), for flying species (vertebrate or invertebrate) such green areas can nevertheless still be functionally connected, with an interchange of species. As we will see in Chapter 2, gardens and other urban green spaces defy the notion that they are barren in terms of species diversity. The challenge is to maintain that biodiversity and hopefully to enhance it. There has been much public concern about wholesale declines of many wildlife species. The worry about the large-scale loss of insect populations has been widely reported in both scientific papers (Sánchez-Bayo and Wyckhuys, 2019) and in the press. The popular term 'insectaggedon' has been coined to emphasize the potentially catastrophic ecological impact of such a continuing trend of insect decline. In reality, however, there are insufficient data to justify the more extreme claims of insect loss on a continent-wide or, in many regions, countrywide scale. Having said that, the UK does have longer-term insect data (up to fifty years of recording) that confirm that insects such as butterflies, moths, bees and hoverflies have significantly declined in abundance. A recent study on cropland areas in the UK confirms the continuing decline in insect biodiversity and abundance, despite changes in agricultural policy and support with greater emphasis on environmental objectives (Mancini et al., 2023). Bearing that in mind, it is good to think of gardens as additional and inviting refuge areas that, with more benign garden management, can collectively contribute to make space for nature.

Learning more about the ecology of a garden helps us understand and appreciate what makes a garden function and how this may lead to a more thoughtful and constructive approach to gardening and garden design, following the science to make gardens work for biodiversity. We will primarily be covering the 'terrestrial' part of the garden. That is to say, there will be little reference to ponds. That isn't to downgrade their importance – it's just that ponds represent a very different ecology; a topic all of its own. Likewise, there will be bias towards the smaller, less visible aspects of the garden ecosystem. Birds, mammals and other groups do get honourable mentions where appropriate, but they are generally already well known and much discussed. Biodiversity encompasses the huge richness of species beyond the 'obvious' groups. These more cryptic species are essential parts of the garden ecosystem jigsaw. Showing the way towards understanding this jigsaw is the purpose of this short book.

Chapter 2
Garden Biodiversity: Endless Variety

If we want to understand the workings of nature in a garden, a good start is an audit of what is there. We began the first chapter with a mention of some of the 'obvious' wildlife in the garden – but we need to dig deeper. It's the organisms that aren't immediately obvious that add to the complex of species that make up our garden 'nature reserve' and create the interactions of a functioning ecosystem. Of course, every garden will be different, but what we need is an insight into the variety of life that will be missed in a cursory familiarity with the wildlife in the garden. The title of this chapter refers to 'endless variety'. It can't truly be endless – but there is likely to be much more than we at first imagine.

We begin by thinking about how to measure that variety. Having said that, one of the first things to note is that, ecologically, gardens are not that well studied. You won't find a large portfolio of scientific papers providing a full audit of what life is found in a particular garden. In the past, ecologists have seen little value in studying gardens. For a start, because gardens are all different, it is difficult to draw broad scientific generalizations from such studies. By contrast, there is a well-developed scientific literature on gardens from the perspective of growing garden plants – the science of horticulture.

However, the relative paucity of detailed scientific studies of garden biodiversity doesn't imply that the task is hopeless. A pioneering book on the subject (Chinery, 1977) described the variety of garden wildlife of all shapes and sizes. It is in part an identification guide (with illustrations) and partly an account of the biology and ecology of various commoner UK garden 'wildlife' species.

Since then, a number of quantitative garden studies have been positively eye-opening in what they have revealed. Some have been carried out by universities, but equally important and valuable has been the contribution of amateur naturalists and local natural-history or nature groups. This is especially true in relation to long-term studies, which are more difficult to run in professional settings where time is money!

The easiest way to think of biodiversity in gardens is to count species. Such a count is expressed using the ecological term 'species richness'. There is scope for confusion here because the terms diversity and biodiversity are used widely in popular conservation literature – but they are not the terms ecologists use to specifically indicate a species count. Nevertheless, even in this book they will be used somewhat interchangeably.

Having pointed to a bit of vagueness in use of terminology, let's begin counting.

The first point to make is that, in practical terms, it is impossible to do a total species count for a garden. This is because we would have to stray into the world of bacteria and viruses. Bearing in mind that the microbial population may vary enormously across the garden, a lot of DNA analysis would be required to enumerate the different microbial forms. The concept of species gets a bit hazy at the microbial level. So, let us make life a little easier at this stage by focusing on larger organisms – but still nothing like the size of a badger. To be a little technical, we will concentrate primarily on multicellular species – so 'above' (in evolutionary terms) the likes of bacteria and simple single-celled animals and plants such as amoebae.

A key study that led the way in terms of investigating the fauna and flora of a single garden was undertaken by Jennifer

Owen in her suburban garden in Leicester (UK) between 1972 and 2001 (Owen, 2010). That meant thirty years of study. Although Jennifer worked as an ecologist, this project could realistically not have been done in a professional setting because it wouldn't fit the standard rubric of a defined funded scientific research project. It required a great deal of her own time as well as voluntary input from over twenty specialist contributors to help with the identification of specimens.

As it turned out, her fairly modest plot proved to be a treasure house of species. Leaving aside the 'obvious' plants, birds, mammals and amphibians (no reptiles were recorded), there were very large numbers of invertebrate species. She recorded flatworms (Turbellaria), slugs and snails (Gastropoda), earthworms (Oligochaeta), leeches (Hirudinea), woodlice (Isopoda), centipedes and millipedes (Chilopoda and Diplopoda), false scorpions (Arachnida), harvestmen (Arachnida), spiders (Arachnida) and insects (Insecta). And the result? A staggering total of 2,673+ species. Around 75 per cent of these were insects, 18 per cent plants and the rest a mix of other invertebrates and vertebrates (mostly birds). The insect groups with the highest species counts were parasitic wasps, beetles, moths and true bugs. Collectively, these represented about 13 per cent of the total UK species count for these four groups added together (Buglife, 2023). These numbers would be way beyond most people's expectations, I think. The full list, taken from Jennifer Owen's study, is shown in Table 1. Of course, this was the grand total over thirty years, so the numbers in any one year would be significantly lower. However, sampling is, by definition, only a partial (incomplete) survey of everything in the garden. Some areas, such as the soil invertebrates, were not systematically sampled. So, it is likely that many species were missed in the yearly totals. But whichever way you take the numbers, they are an impressive testament and riposte to the argument that gardens have little value for wildlife. And even though many of the species were small – and might be considered insignificant – they emphasize the beautiful complexity of garden ecology and ecosystems.

Scientific name	Common/ English name	Species count (n)
Bryophytes	Liverworts and mosses	38
Vascular plants	Ferns, conifers and flowering plants	436
Molluscs	Slugs and snails	17
Annelid worms	Earthworms and allies	5
Platyhelminth worms	Flatworms	4
Arachnids	Spiders and allies	92
Myriapods	Millipedes and centipedes	12
Crustaceans	Woodlice and allies	8
Insects: Odonata	Dragonflies and damselflies	7
Insects: Orthoptera	Grasshoppers and allies	5
Insects: Hemiptera	True bugs	201
Insects: Hymenoptera	Bees and wasps	771
Insects: Coleoptera	Beetles	442
Insects: Diptera	Flies	145
Insects: Lepidoptera	Butterflies	23
Insects: Lepidoptera	Moths	375
Insects: other	Remaining small orders	28
Amphibians	Frogs and toads	3
Birds	Birds	54
Mammals	Mammals	7

Table 1 Jennifer Owen's species totals by groups (taxa) – combined data for 1972-2001 (Source: Owen, 2015).

This study and others that have followed the same model are the result of enthusiasm and passion for natural history as much as for science. One excellent example is a study by Penny Metal (2017), who got interested in the invertebrate wildlife of her small local park in Peckham (UK), a largely residential area not far from central London. The park is described by Penny herself as 'insignificant'. It is definitely small, only 1.52 hectares/3.8 acres, a local amenity park with no pretensions to be a nature reserve. And yet it too produced a surprising tally of invertebrate life. Using her camera as a recording instrument, Penny started her observations in 2011 and in 2017 published a lovely photographic record of what she had found. It is not a scientific text as such, but includes a list, mostly to species, sometimes to genus and occasionally with a more generic label. The total was 555 'species' – a number that gives hope and encouragement to anyone interested in the convenience of recording 'on your doorstep'. And it provides yet further proof that a range of wildlife (invertebrates, in this case) continues to thrive in urban settings. The species may not be rarities, but collectively they do contribute to maintaining an altered, managed and urbanized ecosystem in some kind of quasi-natural state.

There have also been studies involving groups of enthusiasts or professionals collecting data on garden wildlife. We will come to university-led studies, but for the moment we can cover a study that could be described as 'amateur' – albeit with input from a range of people who might still be working in ecology, some retired ecologists and self-taught naturalists. This study had the ambitious aim of documenting nature in the city of Cambridge in the UK (Hill, 2022b). Once the study area had been precisely defined, a variety of approaches were used to collect data for different groups of organisms. One of the collected data sets concerned sixty gardens that were surveyed for 'weeds': all the species that were not intentionally planted by the gardener but were present in flower beds or in lawns. Again, the totals were impressive. There were 341 weed species recorded, of which

204 (60 per cent) were native, 48 (14 per cent) were ancient introductions (archaeophytes – pre-1500) and 89 (26 per cent) more recent introductions (neophytes – post-1500). The average number of such weed species per garden was around twenty-four. So, again, perhaps a surprising number of 'accidental' plants alongside the intentionally planted ornamental ones. Especially as weeds must escape the eagle eye of tidy gardeners!

Since the new millennium, academic interest has also turned to the study of garden ecology. Notable among these studies is the work undertaken by the University of Sheffield in the UK along with a number of associated institutions. This is the 'BUGS' project – Biodiversity in Urban Gardens in Sheffield. The project started with ecological assessment of sixty-one urban gardens. Over the years, this research has involved numerous researchers and spawned well over a dozen scientific papers. It generated a wealth of data on various aspects of garden ecology, conservation and management of gardens. For an overall summary of the initial objectives of the project and a selection of outcomes, see Gaston et al. (2004). We will come back to aspects of the data presented in these papers in due course. Just to give a flavour of the information gathered, one part of the study involved the humble lawn. Although this is thought of as grass, the BUGS study recorded a total of 159 plant species (including grasses) from fifty-two lawns (Thompson et al., 2004). Rather than being a dull monoculture, the average lawn had twenty-four species of plants – the majority of which were native. As gardeners start to become less precious about their lawns and begin to live with species like white clover, even the humble lawn can act as a resource for other wildlife. A regularly cut lawn may still allow low-growing clover to flower and provide nectar for bees. A more recent concept of 'tapestry lawns' is discussed in Chapter 3.

The examples listed from several studies suggest that an inventory of species in any garden will yield much higher counts than might be expected from casual observation. We will refer to particular groups of organisms (taxa) in later chapters, but the next chapter provides more information on the ecological framework of a garden.

Chapter 3
So Many Habitats and Gardens Large and Small

An insect flying over a grassland or through woodland is faced with a mixed but rather uniform sweep of plants, soil, rocks, trees or other features on which to settle. Many will be looking for the right type of habitat to fulfil the needs of surviving the environmental conditions as well as feeding and breeding. Less mobile species must stay local. For instance, a woodlouse (Isopoda) will easily desiccate and has limited mobility; hence leaf litter and rotting logs can serve all these purposes within one small area. By contrast, when flying into a garden our buzzy insect is typically faced with a smorgasbord of choice. Will it be the lawn, the herbaceous border, that overgrown shrubby bit at the back of the garden or maybe even a pond or bog garden? The truth is that the enthusiastic gardener doesn't like uniformity. He or she wants to create a diverse landscape in miniature: the gravel bed perhaps reminiscent of dry Mediterranean flora or the rock garden a European mountainside, creating a niche for a typically 'alpine' plant. There are also more recent innovations in features or planting styles in gardens – all said to promote biodiversity and other benefits. These include planting mini-meadows or incorporating a green roof on a new house extension. As well as providing varied plant and planting interest, this melange of habitats also reflects a degree of artistry and design aspired to by many gardeners – very much part of what gardening is about. While the garden may not turn out quite like Monet's garden at Giverny in France, different zones of the garden provide variety in plants and habitat types for any incoming insect or other species and in that sense should enhance the species count. This would be the 'natural' assumption. Is there any evidence to support this argument?

Smith et al. (2005) reported data from the BUGS project mentioned in the previous chapter. The average size of the sixty-one urban gardens surveyed was 173 square metres/1,862 square feet, probably typical of tightly packed urban environments and close to the average garden size in the UK. On the other hand, Hill's (2022b) study of sixty gardens in Cambridge (see Chapter 2) recorded an average garden size of 466 square metres/5,016 square feet, albeit with a clear trend for reduced garden size in post-1970 housing. Coming back to the BUGS study, this identified twenty-two different kinds of landcover in gardens. These included the usual potential habitats such as grass, flower beds, vegetable patches, compost heaps, ponds, hedges and trees – but also a range of man-made features such as paths, walls, greenhouses, gravel and so on. A typical range of habitats that can be seen in a small garden is shown in Figure 1.

There were large differences in the proportion of the garden allocated to different landcover types. For example, cultivated flower beds varied from around 1 per cent to 59 per cent, while mown lawn ranged from about 1 per cent to 86 per cent. It was perhaps not too surprising that the number of landcover features generally increased relative to the area of the garden – hence the number of potential habitats/resources. In particular, larger gardens were more likely to contain trees taller than 2 metres/6.6 feet, vegetable areas and areas for composting. On that basis alone, a survey of a larger garden might be expected to yield more species. Different groups (taxa) responded in different ways to garden characteristics. However, overall, there was no clear indication of this trend in the BUGS study – at least for invertebrates. The authors suggested that this was probably due to invertebrate mobility allowing movement and mixing of species

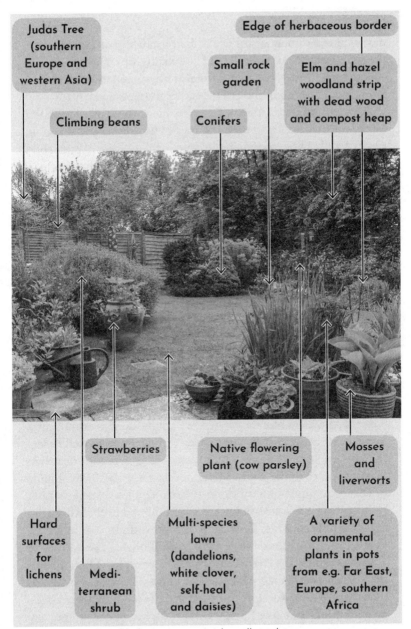

Judas Tree (southern Europe and western Asia)

Edge of herbaceous border

Climbing beans

Small rock garden

Conifers

Elm and hazel woodland strip with dead wood and compost heap

Strawberries

Native flowering plant (cow parsley)

Mosses and liverworts

Hard surfaces for lichens

Mediterranean shrub

Multi-species lawn (dandelions, white clover, self-heal and daisies)

A variety of ornamental plants in pots from e.g. Far East, Europe, southern Africa

Figure 1 A selection of habitats in a typical small garden.

across garden boundaries, which had a homogenizing effect on species distribution (Smith et al., 2006a; 2006b). Free movement of ladybird beetles between gardens was suggested by data in a study in California in the United States (Egerer et al., 2018).

A further interesting approach to check the level of mobility between gardens of four butterfly species was reported by Angold et al. (2006). They examined the genetic relatedness of butterflies at a range of geographic locations across the city of Birmingham (UK). In general, genetic markers showed little indication of isolation of sub-populations of each species. There seemed to be consistent genetic mixing over large distances between urban and peripheral rural populations of the butterflies.

So, in terms of species richness, perhaps garden size doesn't matter thanks to species mobility. The one exception to this was in relation to the distribution of large trees and large shrubs – both of which were more likely to be present in larger and older, more mature gardens (Smith et al., 2005). Because these often support a particular suite of species, tree canopy cover might enhance local species richness and abundance.

What the BUGS study did suggest is that total species counts are likely to reflect the variety of habitats across a range of contiguous gardens rather than individual ones. But with particularly mobile species, we might need to think on a larger scale – because the presence of a species may well reflect the garden habitat composition across larger blocks of urban housing or even beyond. Thus, a relatively rare but mobile species may turn up in a garden even though no suitable habitat or host species is immediately apparent. I get a hint of this from my own garden when I regularly dig up walnuts – brought by grey squirrels – in my flower beds even though I don't know of a walnut tree in the surrounding area of gardens. The squirrels are clearly wandering near and far. There is an advantage to these kinds of exploratory wanderings: the location of resource availability is not fixed and can be patchy; hence it can pay dividends to explore and catch new opportunities. We have already noted that a species like a woodlouse is likely to be more sedentary. However, if the immediate environment gets a little dry, even a woodlouse is able to sense differences in humidity levels and move itself to a wetter location – albeit not very far away.

All this doesn't refute the point that the presence or absence of a species in a particular garden can depend on access to specific resources (for example, food plants) – even if the species ranges over a wider area. We can take the example of moths. These need to lay eggs on plants that will suit the feeding requirements of the larvae. Some moth larvae can feed on many different host plants (polyphagous), and a few on just a single host species (monophagous). An example of the former is the common large yellow underwing (*Noctua pronuba*), whose larvae can feed on many different host plants. At the right time of year these will be readily caught in the UK – sometimes in large numbers – in a moth light trap. On the other hand, the centre-barred sallow (*Atethmia centrago*) moth larvae feed on only a single plant species, namely ash trees (*Fraxinus excelsior*). Because ash trees are common in the UK (notwithstanding the current impact of ash dieback disease), this moth species is also fairly common. I have caught this species in my garden – despite the absence of ash there and in adjacent gardens. But this example also emphasizes the point that many moth species will feed on trees rather than herbaceous vegetation. Larvae of the large and spectacular poplar hawk-moth (*Laothoe populi*) feed on a range of plant species – but they all happen to be trees (plus one or two shrubs). We previously noted that larger gardens are more likely to grow trees – so egg laying by this species and their caterpillars are more likely to be encountered in the vicinity of larger gardens.

Hawk-moths are large and 'showy' moth species, but there are many that are more cryptic. For example, there are a large number of very discrete leaf-mining insects, of which many are moths, while others belong to the order Diptera (flies) and two other insect orders. A variety of herbaceous and woody plants serve as hosts. Leaf miners have larvae that squeeze into and chew their way through leaf tissue within the tiny thickness of a leaf. The evidence for this is a blotchy or sinuous leaf mine – no longer green – where the spongy leaf tissue has been eaten away. The activities of an invasive horse chestnut (*Aesculus hippocastanum*) leaf-mining moth (*Cameraria ohridella*) have become very apparent in recent years but there are several hundred leaf-mining insect species in the UK.

Other moth species are linked to plants found in specific habitats – such as the common reed (*Phragmites australis*) found in wetland areas. If your garden is some way from a wetland, you are less likely to be adding those moths to your garden-species list. Having said that, because the majority of moth species happen to be polyphagous (to varying degrees), most gardens still have the potential to host a large number of moth species.

One other, rather nice example of an association between an insect and a plant is not linked to eating it. This concerns a solitary bee, the European wool carder (*Anthidium manicatum*). The females of these bees are particularly attracted to scraping off the hairs from the rather hairy lambs' ears plant (*Stachys byzantina*), a non-native ornamental often grown in gardens. They use the hairs to line their nests and will also take nectar and pollen from the flowers. Meanwhile, the males become very aggressively territorial around these plants because they are a magnet for females and hence provide a mating opportunity. So, a good way to get the bee is to plant the plant.

Moving on from moths and bees, we can also argue that the resources the gardener provides in the garden may well favour one species rather than another. One example of this is the provision of supplementary food in garden bird feeders. A total of 133 bird species has been reported as taking food from feeders in the UK. There has been a trend towards both an increase in the range of birds adapting to taking food from feeders and the extent to which such feeding has favoured particular species. This is leading to a restructuring of bird communities – with winners and losers, even on a national scale in the UK (Plummer et al., 2019). This study showed a dramatic increase in the use of feeders among a large number of bird species – such as wood pigeon (*Columba palumbus*), goldfinch (*Carduelis carduelis*), long-tailed tit (*Aegithalos caudatus*), magpie (*Pica pica*), great spotted woodpecker (*Dendrocopos major*) and many others. Although the data are quite scattered, there is a general relationship of increased feeder usage leading to positive population increase. Following on from this, the attraction of birds to feeders may then impact numbers of invertebrates in the garden, such as a reduction in ground beetle abundance, presumed to be caused by increased bird predation in the vicinity of bird feeders (Orros et al., 2015). One change causing another.

There can be a variety of overlooked or small habitats in gardens. The range of some of these is shown in Figure 1. One habitat missing from the photo is roof tiles, which provide colonization space for mosses. These lowly plants (in evolutionary terms) hold their own interest – there are over 1,000 species in the UK. But they also harbour small animal life. Particularly fascinating is an obscure group (taxon) of animals called water bears or tardigrades. Place a bit of roof moss in a dish of water, break it up and you are likely to find these fascinating tiny (<1mm/0.04in) eight-legged creatures – but you need a good magnifying glass/loupe or preferably a microscope. Tardigrades are renowned for being able to survive the most extreme environmental conditions (high temperatures, dehydration, radiation and so forth) and still spring back to life. Another lowly plant often seen growing on the surface of damp soil in flowerpots is the liverwort *Marchantia*. Liverworts are grouped together with mosses (in the Bryophyta) but have a flattened growth form (a green thallus) rather than the upright leafy form of mosses. *Marchantia* grows rather charming umbrella-like reproductive structures from the thallus and so is quite distinctive at that stage.

Although the bulk of discussion in this chapter has been about the physical and green infrastructure of the garden, there are environmental factors that will influence species using the garden. We have already had cause to mention aspects like humidity and moisture, but also need to consider shade. In shaded areas, insects are generally less active because of lower air temperatures and an inability to raise their body temperature by direct exposure to sunlight. Consequently, flying insects such as the various kinds of pollinators will be more reluctant to forage and will generally stick to sunny areas of the garden.

A variety of garden habitats and environmental conditions should provide opportunities for many species and a biodiverse garden. The term 'biodiversity' has had increasing traction on the world stage – especially through the Convention on Biodiversity, which has been ratified by all nations except the US and the Vatican! The trickle-down effect of all this attention has been to encourage countries and organizations to adopt the language of biodiversity as something to promote and aspire to in their activities. So, what are the aspirations of the UK's most influential gardening organization, the Royal Horticultural Society? Their recently published *Sustainability Strategy* (RHS, 2021) makes a strong commitment to biodiversity in gardens. To that end, their horticultural specialists have recently been joined by a biologist and a wildlife specialist: recognition by the RHS that gardens can be an important reservoir and refuge for biodiversity – not just 'pests' and 'weeds'.

We will finish this chapter by discussing three innovations that have found their way into gardens and consider whether they have a role in promoting biodiversity.

Green Roofs

We mentioned earlier that a green roof can be a novel addition to a garden. In recent years the enthusiasm for green roofs has really taken off – among architects, town planners and others. Such are their claimed virtues to help propel the greening of cities that they are even being installed on bus-shelter canopies. For a homeowner planning an extension or a garage, adding a green roof can be an option, albeit not necessarily a cheap one. Quite apart from the potential of flat roofs to leak if badly constructed, the weight of soil/substrate used requires sturdy roof support. A key attribute of such a roof is said to be the enhancement of biodiversity. How real is this and is there any evidence that this would make a significant difference to the overall species richness of a garden?

First of all, it is important to recognize that the nature of the substrate on the roof will dictate much of the ecology of that space. Normal soil is replaced with a largely porous substrate low in organic matter – which must be both water absorbent and free draining to avoid waterlogging. The depth of substrate is typically 10–20cm/4–8in. Inevitably this then dictates the kinds of plants that will tolerate difficult environmental conditions. If the roof is not irrigated, plants will (typically) be low and slow growing, hardy, drought-tolerant taxa. In Europe, choice of plant is often limited to an initial planting of native stonecrop (*Sedum acre*) – which is a succulent and naturally found growing in free-draining, drought-prone areas. Of course, a flat-roof area will be subject to a continuous influx of seeds either carried there by air current or by birds. So even if the roof starts with a uniform covering of stonecrop plants, other species will appear. They do so when conditions are right for germination, but a period of drought may mean they do not survive. It is an unstable, rather chaotic plant community. A variety of colonizing species may persist, such as flattened meadow grass (*Poa compressa*) and herb robert (*Geranium robertianum*), both adapted for growth on dry, free-draining substrates.

There has, of course, been more ambition among some city planning authorities and a requirement for more diverse planting. This has necessitated the acceptance of variation in topography on the roof and with that the engineering challenge of greater 'soil' depth and hence a greater variety of species.

Drawing on their natural ecology, species like viper's bugloss (*Echium vulgare*), found on gravelly/sandy soils, and marjoram (*Origanum vulgare*), found on free-draining chalky soils, can thrive alongside the stonecrops. These types of roofs are specifically designed to attract wildlife and can be referred to as 'living roofs'. They require more construction, maybe irrigation, management and expense.

As far as animal species are concerned, the *Sedum* roof does provide a home to many of the usual soil invertebrates – such as mites and springtails – albeit as an impoverished food web with drought-tolerant (xerophilic) species (Thuring and Grant, 2016). When flowering, stonecrops and other plants attract bees and a variety of other insects. But Williams et al. (2014) noted that 'Green roofs largely support generalist species particularly insects, but their conservation value for rare taxa, and other taxonomic groups especially vertebrates, is poorly documented.' Knapp et al. (2019) noted that green roofs 'are not a substitute for' ground-level habitats or green spaces. So as an addition to garden habitats it would be fair to say that a domestic green roof will probably not materially change the species richness that can be achieved in the garden but does increase the area of 'green'.

Mini-Meadows

An ancient meadow, in the UK or other parts of Europe, can be a glorious sight in spring and summer. Having been left 'unimproved' for many years with no herbicides, fertilizers or other chemicals, a meadow will reward with a vivid and varied display of flowers. The absence of chemical inputs allows a wide array of large and small flowering species to find a niche in the meadow. No herbicides will mean a full range of broad-leaved species, while the low fertility of the site prevents dominant species outcompeting others. This will be especially true of grasses, which often respond to the presence of nitrogen in soils and can become rather 'thuggish' in nutrient-rich soils.

Seeing meadows of this type has encouraged gardeners to adopt the idea of turning at least part of their garden into a meadow – a mini-meadow, if you like. This seems like an easy win for biodiversity in the garden. And, of course, it is true – a meadow like this would provide aesthetic pleasure but also a new point of attraction in the garden for a range of insects and other species.

The idea is great, but the execution may not be so straightforward. The situation will be different with an 'annual' meadow versus a 'perennial/biennial' meadow. The annual one is the more straightforward – you clear some ground of other vegetation, sow a suitable annual flower mix (many available commercially) and await a riot of flower colour. To maintain the richness of flowering, you will best repeat the process the following year – albeit there will be some carry-over of species that have set and shed seeds.

More complicated is the establishment of a perennial meadow. Again, there are commercial seed mixes available of perennial meadow plants. But it is the nature of the soil that can prevent successful long-term establishment of a multi-species meadow. Soil that makes life too comfortable for growth (especially soils rich in nutrients) will encourage more vigorous species to out-compete the less 'pushy' ones. As meadow mixes include grasses, it is often the grasses that take over but also more robust perennials. After an early steady increase in flowering (perennials take time to establish) the species richness of the garden meadow may decline. Unfortunately, if you have an intrinsically nutrient-rich soil (clay, for instance), it is hard to correct without some serious earth moving and soil replacement. On the other hand, a more sandy or chalky soil, which gives plants a hard time, can work well with the right perennial seed mix.

Assuming that you are going to go ahead with a garden meadow, what will it add to garden biodiversity? Griffiths-Lee et al. (2022) set out to investigate this through a citizen-science project involving the planting of 150 'mini-meadows' in private gardens (2 × 2 metres/6.6 × 6.6 feet). The plots had three different treatments: two were sown with different seed mixtures (twenty-nine and twenty-four species respectively, mostly perennial species) and the third was a control with no seeding. The study was carried out over two seasons. The results were scored in terms of pollinators (bumblebees, butterflies, moths, honeybees, hoverflies, solitary wasps and 'other' flies/insects). In general, there was little difference in total insect abundance between treatments in Year 1 and Year 2. But there were differences between individual types of insects in Year 2, when there was more flowering. The main positive was a significant increase in abundance and richness of bumblebees, solitary bees and solitary wasps in seeded mini-meadow plots. A third group of beneficial insects, the hoverflies, showed little discrimination between the treatments.

These types of data help to confirm the value of floristic richness for attracting beneficial insects. But it's worth noting that not all flowers will be equally attractive to different insect types. Thus, Campbell et al. (2012) discussed aspects of flower morphology and how it determines insect visits. Bumblebees preferred tubular flowers whereas solitary wasps and hoverflies preferred flowers with shorter, more open petals where nectar was more accessible. The main seed mixes for sale usually contain a range of flower sizes and shapes.

Companion Planting

There is an extensive body of popular literature extolling the virtues of companion planting on vegetable beds (for example, see extensive plant pairings in Walliser, 2020). The argument goes that interplanting secondary plant species alongside the crop has the potential to enhance the growth and productivity of the crop. Typically, claimed benefits relate to reduced pest infestation – which the companion plant may achieve in a variety of ways. For example, it may act as a trap plant, luring the pest away from the crop plant. Or the companion plant is strongly aromatic, which may confuse a pest that seeks out a host plant through its odour – by smelling it (olfaction). For instance, it is known that carrot-root flies 'smell' their way to a host carrot plant.

A study by Finch et al. (2003) tested twenty-four bedding, aromatic and other companion plants that were interplanted with cabbage and onion crops to protect them against cabbage-root fly and onion fly. Some did reduce pest attacks, but this was not a particular property of companion or aromatic plants. Rather, it seemed to be the presence of extra green-leaf area that complicated host finding by the pest species: the companion plants got in the way. So, there is some evidence for why this type of approach could work. The problem is that the body of research to support some of the suggested planting combinations is rather thin. Many of the lists of crops and possible companion plants are based on anecdotal rather than scientific evidence – gardening folklore, in other words. That doesn't mean it should be dismissed – some of the pairings of crop and companion plant will be useful. For example, planting borage (*Borago officinalis*) as a companion plant for strawberries resulted in higher yields and quality. The authors suggested that this was due to enhanced pollin-ation of strawberry flowers by insects attracted to the borage plants (Griffiths-Lee et al., 2020). From the point of view of wildlife gardening, whether it 'works' or not, companion planting will be a plus. The companion plants may provide hosts for additional insects, such as true bugs feeding on sap, while flowers may prove attractive to pollinators and other beneficial insects.

Tapestry Lawns

The term 'tapestry lawn' is a recent reimagining of the traditional garden lawn. Gone is the idea of a pristine swathe of religiously mown fine grasses – essentially a kind of horticultural 'frame' to set off the more exuberant pictorial floral displays in the garden.

Of course, there is no denying that such a lawn can be aesthetically pleasing, but ecologically it can be rather sterile and hard to justify in biodiversity terms. Yet it can occupy a large part of the area of a garden.

However, before the lawn is seen as an ecological abomination, for many gardeners the kind of pristine lawn I have described is a long way from the reality. Many gardens have a lawn that is kept tidy by mowing, but that gradually accumulates a range of broad-leaved species besides the grasses. Without regular use of weedkiller or intensive hand weeding, the shift towards a multi-species lawn is almost inevitable. As this happens, those species that can cope with regular 'cropping' by the lawn mower will often grow with a prostrate habit and manage to flower. It is likely that many UK gardeners will have seen daisies (*Bellis perennis*), dandelions (*Taraxacum officinale*), white clover (*Trifolium repens*), self-heal (*Prunella vulgaris*) and other species all manage to prosper among the grasses despite the mowing, trampling and competition from grasses.

It is possible to buy a lawn-seed mix that includes seeds of wild flowers. But if we go one step further and specifically create a lawn only with mowing-tolerant perennial flowering plants, we have the concept of a tapestry lawn. Many of the plant species suitable for this type of 'lawn' can be described by the technical term 'hemicryptophytes' – that is, plants with buds at ground level and therefore tolerant of mowing. It is argued that such lawns require less maintenance effort, including less frequent mowing (three to five times a year). Like a grass lawn, it should be self-sustaining; but it rewards with a mixed floristic display that in turn is much more attractive to a range of insect species. Experiments with native and non-native mixtures of plants showed that the native-species lawns were more attractive to insects (Smith et al., 2015). Successful trials with establishing tapestry lawns used up to ten to twelve species with no grass. An example of ten native species suitable for such lawns in UK conditions would include the following:

- yarrow (*Achillea millefolium*)
- daisy (*Bellis perennis*)
- mouse-ear hawkweed (*Pilosella officinarum*)
- creeping cinquefoil (*Potentilla reptans*)
- selfheal (*Prunella vulgaris*)
- creeping buttercup (*Ranunculus repens*)
- lesser stitchwort (*Stellaria graminea*)
- white clover (*Trifolium repens*)
- germander speedwell (*Veronica chamaedrys*)
- sweet violet (*Viola odorata*)

Ironically, my own lawn already features several of these, albeit with the bulk of plant biomass still as grass – they just 'arrived' after the lawn was originally planted with turf.

Establishing a tapestry lawn is not straightforward, but ultimately should be rewarding in both aesthetic and biodiversity terms. The practical details of establishing a tapestry lawn, its management and ecology have been described in detail in a book by Lionel Smith, the developer of the concept (Smith, 2019).

Chapter 4
Garden Flora: Native Versus Weeds Versus Aliens

A number of different terms are used to indicate the origin of plants found in a particular country – were they always there or have some arrived more recently? Maybe at some point they were accidentally or deliberately introduced by people travelling the world? These questions have relevance to gardens because they often contain such a mix of species – often from far-flung corners of the Earth. Is it important to know where a plant originates? Yes, because it can have an effect on the ecology of that plant in a garden. The insects and other animals that interact with plants in the garden are mostly local in origin. How do they cope with a 'foreign' plant? We will consider the ecological issues in a later chapter. For the present we need to understand some of the terminology used to describe plants with different origins.

Native

Gardens can contain a huge variety of plant species. Using the term rather loosely and unscientifically, gardens could be thought of as 'biodiversity hotspots'. The plants that feature in gardens may be native wild plants that are also grown as garden plants (for example, the ox-eye daisy *Leucanthemum vulgare*, a common wild flower in the UK but also grown in gardens). What constitutes a native plant is defined by the period of past ice ages. The last of the ice receded in the UK and mainland Europe around 12,000 years ago. There was ice across much of midland and northern areas of the UK, while a good part of southern England remained cold but ice-free during that glaciation. Hence, species considered native are either those that survived the Ice Age(s) towards the 'warmer' south of the country or migrated north to the UK from mainland Europe after the last glaciation. There were periods when the UK had a land-bridge connection to the rest of Europe and when such migration could take place from ice-free regions. This connection with the rest of Europe was severed only about 8,000 years ago, as ice caps melted and sea levels rose.

Of course, it can be hard to get definitive historical evidence of the 'nativeness' of plants, but this can be sought from a variety of sources. There can be early written records going back several centuries, especially of plants that were used by herbalists for the treatment of medical conditions. The first flora of an English county was published for Cambridgeshire in 1660 by John Ray, an outstanding British seventeenth-century natural historian. Going back even further in time, the study of pollen grains (palynology) taken from peat and lake sediment cores can provide direct evidence for the presence of particular species in the UK. Rather conveniently, pollen of different species is very characteristic in appearance and very resistant to decay. On the basis of current knowledge, 1,692 native plants are listed for Britain.

So, deciding what is a native plant is not entirely straightforward; but other definitions get even fuzzier, as we will see in the next section.

Weeds

The term 'weeds' is a tricky one these days. The standard definition of a weed is a plant growing in the wrong place. On this basis it may be one of many native or alien plants that happen to be good opportunistic colonizers of flower beds or lawns. For example, in my garden the native species creeping cinquefoil

(*Potentilla reptans*) has a near-universal presence. For gardeners who like to control what grows in the garden, these random colonizers are a nuisance – both aesthetically and potentially as competitors for species that have been deliberately planted. So why is the term 'weed' tricky? Well, attitudes are changing – or at least there are attempts to re-educate gardeners. There may always be a need for some (eco-friendly) weed control in gardens; but, at the same time, some weeds can be tolerated and can simply be regarded as contributing to garden biodiversity rather than being unwelcome guests. In short, weeds can be a variety of plants of different types, local or from elsewhere – but in some way considered a nuisance, according to this particular gardening definition.

Non-Native/Alien Plants

Plants that don't fit the definition of being native are regarded as non-native. But there are other terms that can be used in this context, such as 'alien'. We can also subdivide the non-natives/aliens based on their time of arrival in the UK or their ecological impact.

The first of these alternative terms is archaeophytes (around 157 species). These are plants that are not thought to be native, but that have been in the UK for centuries (since before around 1500 – the discovery of the Americas) and have effectively blended into the British flora. Very often these species were associated with crops and may have spread through agricultural activity; they are therefore considered as weed species. A striking example is the common poppies (*Papaver rhoeas*) that are still seen en masse in UK cornfields. Furthermore, there are other non-native species that appear in the British flora after 1500 and are referred to by botanists as neophytes (around 1,596 species). Again, many have integrated with the UK flora. Both archaeophytes and neophytes may also be referred to as naturalized when they are integrated with UK flora.

So far so good – but there are extra bits of terminology that also need a mention. We have the term 'alien' in the section heading and this can also be equated with the term 'introduced'. So, both archaeophytes and neophytes can be placed under either of these headings – or referred to as non-natives. In horticultural circles, non-native plants grown for show (and sometimes escaping in the wild, such as South African montbretia (*Crocosmia* × *crocosmiiflora*)) can be referred to as 'exotic'. Another term used in horticulture is 'ornamental'. This refers to plants grown for aesthetic reasons. They can be non-native or showy varieties of native plants (for example, the UK native plant known as sea pink or thrift (*Armeria maritima*), grown in rock gardens).

The last important term I need to mention is 'invasive'. Some plants that arrive don't settle in 'quietly' and largely unnoticed. These are often species that arrive with three advantages. They find the environmental conditions to their liking; they are no longer encumbered by their usual competing plant species; and they have also escaped many of the usual plant eaters (herbivores) and diseases (pathogens) that occurred in their original location. Hence, they do very well and can become a problem both economically and ecologically. How do such plants cross seas and mountains to arrive in the UK? The answer is in lots of ways. They may be transported either accidentally or deliberately by humans or as mobile seeds, which can be carried a long way by wind or water. Some plants have been introduced from abroad into gardens or botanic gardens for aesthetic or interest reasons. Although any garden can be thought of as a kind of botanical 'zoo', you cannot cage plants or their seeds. Given the right growing conditions in their new home, they can 'hop over the wall' and spread. One such example is Himalayan balsam (*Impatiens gladulifera*), which was originally introduced into UK gardens in the nineteenth century as a pretty pink-flowered ornamental. Despite being an annual, it grows ferociously fast – up to 2.5 metres/ 8.2 feet tall in a single season – and forms dense thickets that smother other low-growing native plants. It is now widespread and has the official designation of an invasive species.

As climate change brings ecological challenges, there are some bizarre examples of invasive species establishing in unexpected places. A recently reported example carried a headline in the *Guardian* newspaper: 'Cacti replacing snow on Swiss mountainside due to global heating'. And yes, the story is about the prickly pear (*Opuntia*) from semi-arid regions of the Americas, as an invasive plant in parts of Switzerland (Perrone, 2023).

So, to recap the mix of terminologies we have encountered. The fundamental distinction is between those plants considered native and those that are not. The latter can be described as non-native, alien, exotic, introduced or invasive. The context in which these terms are used can vary, but they all refer to non-native plants. The term 'ornamental' also typically refers to non-native plants, but could be showy forms of native species. Finally, botanists use the terms archaeophyte and neophyte for non-native species that have largely integrated into the British flora. These can also be described as naturalized. Hopefully all that makes sense!

Chapter 5
Plants and Insects: Conflict and Coexistence

We will deal in more detail with the role of garden plants in wildlife ecology later, in Chapter 11. For the present, a few general comments about why all plants are not the same in relation to garden ecology.

Plants of different species are both very similar and different. They are the same in terms of their basic structure – their cells and tissues – albeit that some are woody and some are not. They also have more or less the same basic cellular metabolism, with a range of molecules (primary metabolites) driving biochemical processes. The two processes common to (almost) all plants are cellular respiration (breakdown of sugars made in photosynthesis to release energy to drive other cellular processes) and photosynthesis (the capture of light energy and energy storage in sugars made from the raw materials of water and carbon dioxide). But that is not the end of the story. Plants also contain a variety of so-called secondary metabolite molecules that vary among plant families, genera and species. They can be present both in the above-ground parts of the plant and in the root system. These chemicals can make a plant more attractive (for example, by way of scents) or less attractive (through poisons, repellents, latex and so on) to animal species. Hence, a plant with a battery of defensive chemicals may be avoided by many plant-feeding species (herbivores). Of course, if the plant is native then there would have been a long period of coevolution between these native plants and local animals – leading to some kind of uneasy evolutionary coexistence. And some species may have evolved the capacity to tolerate the otherwise challenging chemical defences of a particular plant species that other herbivores avoid. Even where this happens, natural selection will continue to favour plants that can stay one step ahead of their herbivores. If a plant is non-native but after arrival in a new land is exposed to a native fauna, no such process of coadaptation would have occurred. The alien plant may have no herbivores able to feed on them. This is often a reason why non-native plants become invasive.

It is worth noting that anti-herbivore defences aren't necessarily just chemical. There can be effective structural adaptations – such as hairy stems to discourage aphid (Hemiptera) feeding. These plant hairs may secrete sticky resins or sting – for example, stinging hairs in the case of nettles (*Urtica dioica*). Large spines can be effective against larger herbivores. And some herbivores will be discouraged from chewing through bark or tough fibrous leaves (described as sclerophyllous leaves) or put off by accumulations of hard mineral silica crystals or needle-like calcium oxalate crystals in the tissues.

For insects and other species exploiting garden plants, additional problems arise in gardens that typically include a large number of plants 'tweaked' by plant breeders. Artificial selection of desirable characteristics has made certain flower features more pronounced in ornamental plant varieties – sometimes to the detriment of interactions with animal species associated with the 'normal' form of the plant. The result may be changes in the structure of the plant – foliage or flowers – or even in aspects of the plant biochemistry or physiology. Taking the example of flowers, there may be changes in scent, flower colour and/or colour patterns, or physiological changes such as to the flowering period. All of these can particularly affect pollinator species. One of the more drastic changes often favoured by breeders of ornamental plants is to select for so-called 'double' flowers. Instead of the new plant variety having the normal number of flower petals, let's say four or five, there may be an additional set or sets of petals. This can make the flower look 'showy' – but it may also make the plant unavailable to pollinating insects. The obvious explanation for this is that such mass of petals limits access by insects, but there is more to it than that. To explain, we need to delve a little into flower genetics to understand the origin of the extra petals.

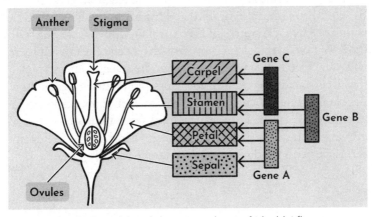

Figure 2 The ABC model and the genetic basis of 'double' flowers.

If we think back to basic flower structure, a typical flower will have a set of green sepals and colourful petals surrounding the reproductive parts of the plant. Usually (but not always), there will be a set of 'male' structures – the stamens (with anthers that release the pollen) – and the female ovule-containing carpel, part of which is a pollen-receptive stigma. The ovules need to be fertilized by the pollen to start the growth of a seed. Normally (but again not always), the sticky tips of the stigma capture pollen from a different plant of the same species, carried there by pollinating insects (or wind). Insects visit the plant not to be helpful but to collect a feeding reward of pollen from anthers and/or nectar – the latter typically secreted from nectaries at the base of petals. So far, so straightforward. But, as shown in Figure 2, the formation of these flower structures during development is under the control of three sets of genes. This is known as the ABC model of flower development. Paired combinations of those genes trigger the formation of sepals or petals or stamens or carpels. So, we can see that mutations that upset the function of gene C will result in a flower with no carpels or stamens but with extra petals in place of stamens – a double flower. There are other configurations possible with mutations in genes A and B, helping to explain some of the oddities that can be selected by plant breeders. Of course, if there are no stamens, not only will the flower be difficult to access through a mass of petals but there will also be no pollen reward for the visiting insect and no pollen to carry to another flower for cross-pollination. In addition, double flowers may have little or no nectar in some plant species (Corbet et al., 2001). The message from this is that double flowers are less likely to be of interest to pollinating insects. They are therefore not a good choice for the insect-friendly garden.

Chapter 6
Plants, Animals and Garden Ecosystems

In the opening chapter, I referred to the garden as an eco-system – but what exactly is an ecosystem? Let's approach this question from the perspective of a somewhat florid description that conjures up the idea of an ecosystem. It comes from the writings of that master naturalist Charles Darwin and is the last paragraph of the 6th edition of *On the Origin of Species* (1872). 'It is interesting to contemplate a tangled bank, clothed with many plants of many kinds, with birds singing on the bushes, with various insects flitting about, and with worms crawling through the damp earth, and to reflect that these elaborately constructed forms, so different from each other, and dependent upon each other in so complex a manner.'

I have already discussed the variety of species to be found in a typical garden. None of these species exists in isolation. As Darwin notes, a complex web of interactions exists between them – which links species across the range of habitats in the garden (referred to in Chapter 3). Not only that, but the mobility of many species will create additional connections beyond the boundaries of individual gardens. Nevertheless, if we describe an ecosystem as a partially self-contained/isolated ecological unit, then there is convenience in considering a garden as an ecosystem. We know that there can be big differences in the design of adjacent gardens and their management, so to that extent each garden does represent a semi-independent unit with a particular suite of resident/semi-resident species. As well as species linking to each other in a variety of interactions, the garden ecosystem will be structured by variations in environmental conditions in different parts of the garden. Coming back to our earlier woodlouse example, the damp conditions shaping the distribution of woodlice in the garden will also shape the distribution of a woodlouse-eating spider as well as fungi and other species that will colonize the damp, decaying log shared with the woodlice. None of these species exists in isolation.

Figure 3 shows an extremely simplified model of how species in an ecosystem interact, which can happen in a variety of ways. It may be a predator–prey relationship (such as the woodlice and spiders) or one species parasitizing another (parasite) or one species causing disease in another (pathogen). Even woodlice have a share of both of the last two – bacterial pathogens and 'parasitic' (parasitoid) fly larvae. Some of the links between species are truly weird and wonderful. Examples are discussed here and in Chapter 10. Figure 3 illustrates only straightforward feeding relationships and includes fewer than twenty species. But it does emphasize an important aspect of ecosystem organization, with feeding relationships creating a food web and so structuring the living part of an ecosystem. In Chapter 2, I gave some examples of garden studies that have included counts of species richness. This reminds us that our food web, were we to incorporate parasitic and pathogenic organisms, would have to have hundreds and most likely thousands or even tens of thousands of links if we included single-celled Protozoa, fungi and bacteria and even viruses. In practice such a food web would currently be impossible to create because, however whizzy computer-generated graphics can be, there are many knowledge gaps – particularly in relation to the microbial world. Nevertheless, it is important to remember that feeding relationships referred to in later chapters all contribute to the complexity of this food web. It is also important to remember that diagrams such as Figure 3 give a static representation. In reality, ecosystems are dynamic, with a constant flow of energy and materials between species.

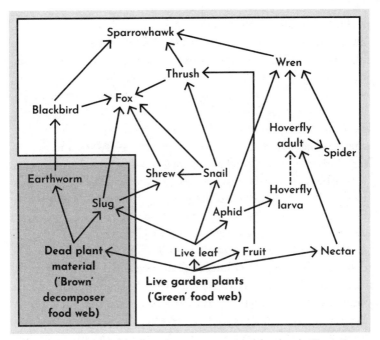

Figure 3 A much-simplified garden ecosystem and food web. 'Brown' refers to the decomposition of dead plant material and the decomposer community. 'Green' refers to the grazing of herbivores on live plant material. In reality, much more plant material passes through the decomposer community than is consumed by herbivores. These two parts of the ecosystem overlap and function together (adapted from Thompson, Wildlife Gardening Forum: www.wlgf.org/food_webs.html).

We already know that any garden ecosystem will not be naturally assembled but subject to regular human intervention through garden design and garden management. Aphids that might have fallen prey to hoverfly larvae are frequently eliminated by pesticides. Most of these pesticides have proved to be non-selective, killing other insects such as hoverfly larvae as well as aphids.

Because organisms are linked by feeding relationships, a particular species may only be present in a garden if it can connect with the next link in the food web – a food source or even multiple sources. This is complicated further when species have different needs for feeding and for breeding. Thus, a cabbage-white butterfly (Pieridae) needs flowers for feeding on nectar, but plants of the cabbage family (Brassicaceae) for egg laying and caterpillar feeding. A stranger example relates to parasitic bee-flies (Bombyllidae), which are often seen in gardens hovering expertly in front of flowers while feeding on pollen and nectar. The dark-edged bee-fly (*Bombylius major*) parasitizes *Adrena* mining bees, which dig soil burrows to accommodate their eggs. Female bee-flies hover in front of mining-bee burrows and flick eggs into the opening! Hence a 'happy' female bee-fly needs both flowers and mining bees within reach to complete its life cycle.

We have already mentioned in Chapter 3 that something as seemingly innocuous as having bird feeders in the garden can affect both the abundance of ground beetles and bird species populations there (Orros et al., 2015). In that sense, these partly man-made garden ecosystems are likely to be only partially integrated and functional. For example, if we focus on the soil component of the garden ecosystem – say, in a vegetable patch – an enthusiastic gardener will dig over the patch each year before resowing/planting. It isn't difficult to imagine how the upheaval caused by digging might disrupt the 'tangled web' of interactions found in a stable soil environment. And these types of ecosystem-disruptive events are inevitably the hallmark of many gardening activities. It is unrealistic to think that a garden can ever fully achieve some kind of 'balance of nature' – it may be permanently in transition. Hence a recent plea from the UK Royal Horticultural Society, as expressed in their sustainability document (RHS, 2021). To quote:

We're encouraging gardeners to recognize their gardens as ecosystems, and breaking down the ideal of 'perfection'. If you can relax some of your practices and take a moment to think about the impact you might be having, your garden becomes much more eco-friendly. Tolerate the wildlife, including pests, and hold your nerve, balance will be restored. A clump of aphids on your rose bush may well become a vital meal for baby blue tits – but only if you don't spray them off first!

Chapter 7
Ecosystem Services in Gardens

Ecosystem services may be on the fringes of familiarity for many gardeners, but these services have increasingly come to the fore as a concept since the United Nations published the *Millenium Ecosystem Assessment* (2005). This UN report had the ambitious goal of a global assessment of human impacts on the environment. In particular, it emphasized the idea of nature as natural capital that is crucial to many aspects of life on Earth, and that it needs to be conserved. It is this natural capital that generates a range of ecosystem services that we often either dismiss or undervalue – at least until things start to go wrong. In a variety of ways, nature and natural processes mitigate some of the harm that humanity does to the planet as well as providing resources that we exploit and other benefits such as pollination and cultural benefits. The variety of benefits which natural ecosystems contribute are collectively referred to as ecosystem services and Figure 4 summarizes the key elements of these services. Some of these can be recognized as relevant to gardens. Although biodiversity is not explicitly listed, it is the fundamental basis of ecosystems and their functioning, but not in itself either goods or services. So indirectly biodiversity contributes to several of the services listed.

Because many of these natural processes and benefits are 'invisible' or not recognized (for example, the self-purification of rivers and lakes after discharges of organic waste), they are often not valued. Economists have argued that ascribing a monetary value to these natural processes is one way in which nature and natural capital are more likely to be seen as a tangible resource. In other words, something that is valuable and not to be destroyed through ignorance, overuse or over-exploitation.

CULTURAL services are the non-material benefits people obtain from ecosystems.

PROVISIONING services are the products obtained from ecosystems.

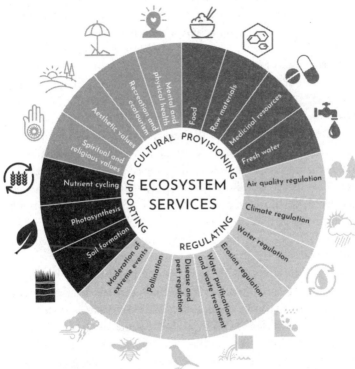

SUPPORTING services are those services that are necessary for the production of all other ecosystem services.

REGULATING services are the benefits obtained from the regulation of ecosystem processes.

Figure 4 A summary of different ecosystem services as defined in the *Millenium Ecosystem Assessment* (adapted from WWF, 2018).

What kind of ecosystem services do gardens provide? The categories in Figure 4 do not necessarily translate directly to a garden scenario in a way that can either be physically measured or evaluated by the gardener or other residents. Because of this, analysis must be partial and focus on what can be assessed. Two scientific studies show different approaches to this type of assessment. The first (Camps-Calvet et al., 2016) was conducted in Barcelona (Spain) and used structured-interview techniques to ask gardeners to identify the benefits that they perceived from their gardens. The researchers subsequently categorized the responses under the heading of different ecosystem services (as identified in Figure 4), namely 'Provisioning', 'Regulating', 'Habitat or Supporting Services' and 'Cultural Services'. The resulting data are shown in Figure 5. Interestingly, by far the most common responses related to 'Cultural Services'. Food supply also scored highly. Perhaps a little disappointing from an ecological perspective, the value of gardens for biodiversity and pollination – closely linked services – were rated relatively low. Note that equally surprising was the high rating of soil fertility – perhaps recognizing the old gardener's adage that 'The answer lies in the soil.' Additional items could have been added to this list – for example, noise reduction, rainwater infiltration and drainage.

The Barcelona study provides a partial introduction to the range of ecosystem services that gardens might provide. However, the relative values of each ecosystem service might well vary in different surveys conducted in other countries – with different gardening traditions, attitudes to and knowledge of garden ecology and other ecological aspects such as climate change. For example, a study in Sheffield on the psychological benefits of using publicly accessible green spaces showed that the user experience was much enhanced by greater biodiversity (Fuller et al., 2007). In relation to gardens, de Bell et al. (2020) summarized the results of their study on the value of gardens in England in the following way: 'Respondents

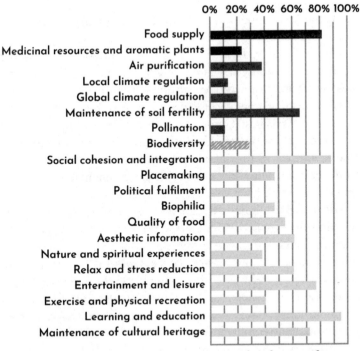

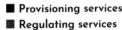

Figure 5 A survey of the perceived value of selected
ecosystem services based on interviews with 44 gardeners
in Barcelona, Spain (Camps-Calvet et al., 2016).

who reported both gardening and using the garden to relax
reported better health and wellbeing, more physical activity,
and more nature visits than those who did not.' Clearly, a
connection with gardens increased the interest of respondents
in biodiversity and nature in general. A more recent UK
study tested the impact of involvement in citizen-science
nature projects and other nature-connectedness activities. All

enhanced well-being and attitudes to engagement with nature (Pocock et al., 2023).

These studies provide some support for the notion of managing green spaces – and gardens – to encourage greater biodiversity, thus providing aesthetic appreciation, well-being and interest. There is also anecdotal evidence of this interest in biodiversity in gardens – as demonstrated by the surge in feeding of garden birds and creating bug hotels and woodpiles for insects and other species. On the botanical side, there is the passion for planting many different kinds of plants – both native and ornamental.

A second approach to gauging the value of ecosystem services is to collect data and estimate specific garden ecosystem properties that can be equated with an ecosystem service. Tratalos et al. (2007) considered water run-off (lack of water infiltration in urban environments that can lead to flooding), carbon sequestration (for example, carbon storage in long-lived trees or organic material in soil) and temperature (the extent to which gardens and green spaces ameliorate raised temperature in urban environments – influenced by the 'heat island' effect). 'Heat island' refers to elevated temperature in cities, compared with surrounding areas, linked to heat absorption and release by roads and buildings and intensive energy use and release of waste heat.

The researchers collected data from five cities in the UK. The key message of this study was that the value of these ecosystem services declined as the density of housing/buildings in urban areas increased. On the basis of the available data for measuring urbanization, address density showed the strongest link to changes in these three ecosystem services. This is shown in Figure 6 (a, b and c). For the purpose of showing trends, the curves here were fitted 'by eye', masking a lot of scatter in the original study data points. This scatter was to be expected with the variations in garden and open-space design. Gardens with extensive paving, for example, will inevitably have a higher value for run-off of rainwater compared with those with

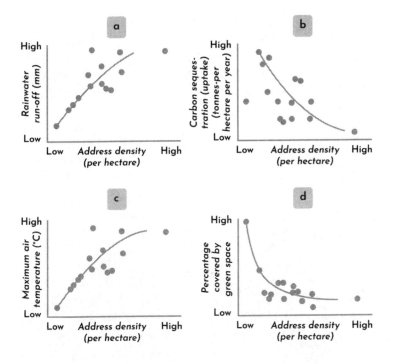

Figure 6 Relationship between the density of addresses and (a) water run-off, (b) carbon sequestration, (c) air temperature, and (d) percentage area covered by green space in fifteen garden study sites across the UK. Curves fitted here 'by eye' to original data point graphs in Tratalos et al. (2007).

extensive grass and flower beds, and hence a poor rating for this ecosystem service. In general, a lower address density in urban areas equated with an enhancement (a higher rating) for the three ecosystem services. The same study also measured green space (Figure 6d), which declined sharply as address density increased – as would be expected. This is likely to impact urban biodiversity through reduced connectivity and habitat variety, as garden sizes get smaller in inner cities.

While the contribution of individual gardens to ecosystem services may be small, collectively they can make a difference.

For example, there has been encouragement from the UK Royal Horticultural Society for a reduction in hard surfaces in gardens – especially in front gardens and patio areas. Even where there is a need for a hard surface to park a car, paving or other hard surfaces can be partial and interspersed with gravel

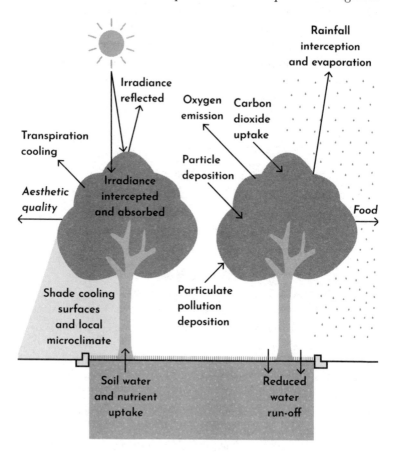

Figure 7. Ecosystem services (see Figure 4) that can be provided by garden trees include: REGULATING – air quality, climate regulation, water regulation; SUPPORTING – photosynthesis; PROVISIONING – food; CULTURAL – aesthetic values (adapted from Livesley et al., 2016).

and low-growing plants such as thyme (*Thymus* spp.) which tolerate trampling. More infiltration of water into soils should reduce the incidence of flash flooding in towns and cities, so is an important potential ecosystem service provided by gardens.

We can finish this chapter by considering the contribution of an individual plant to ecosystem services – in this case a garden tree. If you refer back to Figure 4 for the list of ecosystem services, this then connects to Figure 7, in which a tree is shown contributing to several of those services. We have previously also noted that trees contribute to garden biodiversity, an implicit rather than explicit ecosystem service.

Chapter 8
Gardens and Species Recording

Contrary to common perception, urban areas represent only about 3 per cent of the Earth's land area. Having said that, there are many countries where urbanization is more extensive. In the UK, the overall equivalent value is 6.8 per cent – although it varies from 10.6 per cent for England to 1.9 per cent for Scotland (UK National Ecosystem Assessment, 2011). It is generally considered that urbanizing an area and the loss of natural/semi-natural habitats will be very destructive to wildlife – signifying the end of nature. In this chapter we will examine the extent to which that is true. Can nature survive the ravages of housebuilding and urbanization, or can it still find a home in towns and cities? If so, will the animal and plant community be just a poorer – watered-down – version of the habitats and communities that have previously been present in the area? What kind of species survive in an urban setting – do some species adapt? Given the right kind of management, is there still space for rarer species to treat gardens as a suitable habitat? Accepting that the science of conservation lays particular emphasis on rare and endangered species struggling to survive in a changing world, is the term conservation at all applicable to a garden environment? There are lots of questions – and we need to look for some answers.

During the infamous Covid lockdowns in the UK (2020–21), I started moth trapping in my garden. I was soon amazed by the variety of moths I caught. It wasn't difficult (apart from problems of identification) to reach a list of about two hundred moth species. Most of these I had never previously seen in the garden – even the large and spectacular hawk-moths (also known as sphinx moths). It was heartening to achieve such a tally of moth species in a garden not far from the busy city of Cambridge (UK) and on the edge of an intensively cropped agricultural landscape. It served to emphasize the point that while gardens will have some species that are visible and well known, there are potentially a great number of more cryptic species that will largely be unfamiliar to the gardener. It is this story, of both the more visible and especially the cryptic species, that we will try to unravel in this chapter.

As a rule of thumb, we need to bear in mind that, generally, the smaller the organisms, the less well known they will be. For example, insects are better known than, say, mites (Acari – relatives of spiders). This is because insects are generally much larger, more visible and more readily identifiable than mites – so they generate more interest and are studied more frequently. Inevitably, there are then more experts in insects (entomologists) than in mites (acarologists). So, it can be hard to get help with identification if you develop a passion for mites! When we get to the microbial level the problem is not just technical expertise in identification, but also the ballooning of 'species' number – even if you can be sure what constitutes a species at that level. DNA 'fingerprinting' has both simplified aspects of microbial identification and emphasized the difficulties of drawing boundaries between species and the sheer number of potential microbial 'species'.

We hinted at the question of how we apply the term 'conservation' to gardens. The term can be thought of in the narrow sense, as applying primarily to species that are in some way 'special', or in a broader sense, where it has more of an ecosystem/habitat emphasis and where all species are considered worthy. Small aphids, spiders and flies may not be appreciated by gardeners but provide food for birds and so are an integral part of the web of feeding interactions in the garden ecosystem. This dichotomy has long been a point of debate in conservation – particular species (for example, in the UK the large blue butterfly *Phengaris arion* or the turtle dove *Streptopelia turtur*) have a lot of appeal and generate a lot of public support. They are an easier 'sell' than the more anonymous plants and animal species that are collectively the bedrock of a functioning ecosystem. However, it would be fair to say that in recent years the argument has generally shifted towards a more holistic view (where possible) of conserving habitats – beyond the emphasis just on 'flagship' species.

We will adopt this holistic approach and consider diversity and conservation in gardens in their widest sense. To do this, we can start with the study by Owen (2010; 2015) – see Chapter 2 – which not only has a high species tally but was also carried out over a longer time period (thirty years) than any other garden study in the UK or elsewhere. The Owen garden covered a relatively large area of 741 square metres/7,976 square feet. The long duration of recording has allowed some analysis of how species presence, richness and abundance has changed over time. Table 2 shows a list of the main groups (taxa) recorded in the garden. The total count is an impressive 2,673 species. Beyond that total number, we can look at how species are distributed between the different taxa – but with a degree of caution in interpretation. Why? Because this kind of study invariably requires a good deal of identification expertise. While it's possible for one person to do much of this with a lot of time, patience and effort, expert help with particular taxa will be necessary. The availability of that help is likely to skew the species counts in terms of there being more species where more assistance is available. In this case, the high species count for bees, wasps and ants (Hymenoptera) (771, 29 per cent of the total species count) is real, but probably overestimates the contribution of this taxon to the total species list for the garden because of greater identification effort. This example emphasizes that with additional sampling and expertise most of the taxa totals would be higher — especially for those groups that are cryptic, hard to sample or left out because of identification difficulties. It is also worth noting that the very high count of vascular plants included planted ornamental species — which varied over thirtyyears. These species were excluded in other, similar garden studies, such as those referred to below.

Another equally large data set of species from a smaller garden (200 square metres/2,153 square feet) has been collected in Norfolk (eastern England) (Hodge, 2022), about 225 kilometres/140 miles east of Leicester. This garden is located on the coast. Over a nine-year period, Hodge recorded a very impressive 2,751 species, with a much higher count of invertebrates than in Owen's study, despite the reduced size of the garden and the shorter recording period.

A further shorter garden study has been carried out by Paul Rule in Cambridge (UK), about 115 kilometres/72 miles southeast of Leicester. This was a less intensive study than the other two but still produced a high count of 783 species after about five years of recording (Hill, 2022b).

In each of these three studies, data were collected for a different suite of taxonomic groups, but the data can be compared by looking only at those taxa listed by Owen. What do their results show? First of all, total plant species counts were very different, depending on whether planted ornamental species were included (Owen — 436 species) or were not (Hodge — 158 species). We will come back to the value of ornamental versus native plant species for wildlife at a later stage. But we can draw on another study, the BUGS study in Sheffield, to get a more general picture of plant species in sixty-one urban gardens (Smith et al, 2006c). The combined species count for all gardens was 1,166 species, 30 per cent native and 70 per cent non-native. Because many of the non-natives were rare — often occurring only once — gardens on average actually contained 45 per cent native species. Larger gardens had higher species counts — doubling the garden size led to around 25 per cent more species being recorded. The average number of plant species per garden was 112 (range 41–264), which is surprisingly high bearing in mind that the gardens were surveyed only once and did not include ornamental species. There is a lot of 'hidden' plant diversity in domestic gardens. Furthermore, in any survey some types of plant receive more attention than others – especially herbaceous species, trees and shrubs. What are referred to as 'lower plants' tend to get much less of a look-in. Of these, mosses and liverworts (Bryophyta) are the most likely to be noted, followed by lichens. Algae are the most likely to be ignored but will be rare in a non-aquatic environment. For example, the BUGS study recorded a total of 67 bryophyte and 77 lichen species (Smith et al, 2010). The latter are composite organisms with fungal tissue harbouring symbiotic algae and/or photosynthetic cyanobacteria. Their relevance to the ecology of the garden is confirmed by the large number of moth species with larvae and other small invertebrates that feed on these two groups of organisms. It has been estimated that lichens are the dominant life form on around 8 per cent of the Earth's land surface – emphasizing their global importance despite a low 'public' profile (Asplund and Wardle, 2017).

The total number of bird species recorded in a garden will vary depending on whether you count birds physically present in the garden or add species seen in adjacent gardens or overflying the garden. Hodge's count (140) was higher than that of Owen (54) because it would have included many overflying coastal birds not seen in Leicester. A very clear effect of location.

That leaves us primarily with the invertebrates. There were a lot of big numbers for some of the taxa. This is clear in Owen's data in Table 2. The highest species counts in all three featured data sets (Table 3) were for the true bugs, bees and wasps, beetles, flies and moths, so there is a degree of consistency between the studies. However, if we rank the species counts for the top three groups in each of the three data sets (Owen, Hodge and Rule), they are not the same.

Scientific name	Common/English name	Species count (n)	Percentage (%)
Bryophytes	Liverworts and mosses	38	1.4
Vascular plants	Vascular plants	436	16.3
Molluscs	Slugs and snails	17	0.6
Annelid worms	Earthworms and allies	5	0.2
Platyhelminth worms	Flatworms	4	0.1
Arachnids	Spiders and allies	92	3.4
Myriapods	Millipedes and centipedes	12	0.4
Crustaceans	Woodlice and allies	8	0.3
Insects: Odonata	Dragonflies and damselflies	7	0.3
Insects: Orthoptera	Grasshoppers and crickets	5	0.2
Insects: Hemiptera	True bugs	201	7.5
Insects: Hymenoptera	Bees, wasps and ants	771	28.8
Insects: Coleoptera	Beetles	442	16.5
Insects: Diptera	Flies	145	5.4
Insects: Lepidoptera	Butterflies	23	0.9

Insects: Lepidoptera	Moths	375	14.0
Insects: other	Other taxa	28	1.0
Amphibians	Frogs and newts	3	0.1
Birds	Birds	54	2.0
Mammals	Mammals	7	0.3
TOTALS		2,673	100.0

Table 2 Total species counts for a wide range of groups (taxa) recorded in a 30-year study (1972-2001) in Jennifer Owen's Leicester garden (Owen, 2010; 2015).

Scientific name	English/ common name	Species ranking J. Owen	Species ranking T. Hodge	Species ranking P. Rule
Insects: Hemiptera	True bugs			3
Insects: Hymenoptera	Bees, wasps and ants	1		
Insects: Coleoptera	Beetles	2	3	2
Insects: Diptera	Flies		2	
Insects: Lepidoptera	Moths (only)	3	1	1

Table 3 Comparison of the top three rankings in three long-term surveys of garden insect diversity (Owen, Hodge and Rule).

Why is the ranking different? Well, this could be for a number of reasons. The obvious one is that the different garden environments, resources and locations are not the same – hence they might favour one group rather than another. This may well have some effect on species counts, but there are more probable explanations. In many cases you can't identify invertebrate species just by observing them *in situ* in the garden. They need to be caught for closer examination. But then the species you catch will invariably reflect the trapping methods you use. Every trapping method is selective and may have a strong bias towards a particular group. If you use a light trap at night, then you will catch lots of moths (along with some other night-flying insect species) attracted to the light. If you use so-called pitfall traps – at their simplest, plastic cups sunk into the soil to catch invertebrates moving across the soil surface – you will catch many night-active soil-surface-dwelling ground beetles that fall in. If you use a fine mesh net to sweep through garden vegetation, you are likely to get a good catch of plant-feeding true bugs. This means that if you don't use a wide range of trapping methods in the same way, then some groups will be favoured and not others. The selection of trapping methods used may also depend on available identification expertise. For example, it is easier to identify larger moths from field guides (photos and artwork), whereas beetle identification frequently requires more complex identification keys using technical terminology and often needing a microscope to spot patterns of hairs (setae) and pores on the surface of the exoskeleton. Hence there is a mix of explanations for the differences in ranking between the data sets.

These species totals are truly impressive for gardens that are often so altered that they seem to show little connection to natural habitats. But are these species counts the full story? The truth is that it is hard to tell how many species have been missed and so what the real totals might be. But it is possible to tell from the data whether many more species are likely to be found if there is more sample collection. A simple garden example would work as follows. Let's say that there has been ten years of collection of data on ground beetles in a garden. The data are presented as a cumulative species count – the total number of new species in each year is added to all the species caught up to the previous year and so on. So, there are relatively few species in year one – but a large species total by year ten. Figure 8 shows these hypothetical data plotted on a graph, with two different data sets as examples. By year ten, the total species count in Garden A is no longer increasing much – whereas with Garden B the curve continues to rise. In ecological jargon these are known as rarefaction curves. At least in the case of Garden A, the data suggest we have captured most of the species that are present. With Garden B there is still a fair way to go. Jennifer Owen herself implied that her garden would probably be closer to curve B – with the suggestion that the total number of insect species in her garden could exceed 10,000 by continuing recording efforts (Owen, 2010).

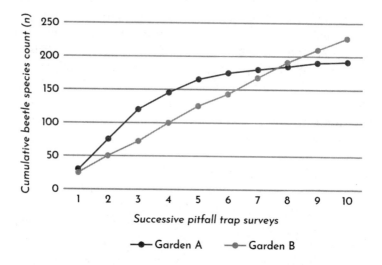

Figure 8 Hypothetical cumulative species counts for ground beetles in two gardens: rarefaction curves.

Can we make any comment about the 'quality' of the species recorded in gardens? After all, in the world of conservation there is often an emphasis on 'rare' species. Owen (2010) divided the records of the larger moths she caught (macro-lepidoptera) into four categories, namely common, uncommon, scarce and 'recorded once only'. The last list included around fifty species caught only once in thirty years. This doesn't directly say anything about their national status in terms of rarity – they may just be locally uncommon – but some at least are likely to be regionally or nationally rare. Britain has a great record of naturalists collecting data on a wide range of species, many of them very knowledgeable amateurs. Data of this kind – even for currently common species – provide a wealth of evidence for filing gaps in distribution maps and analysing long-term changes in fauna and flora. This is particularly pertinent at a time where there is much concern about climate change. The second point is that having the 'eyes and ears' of this army of naturalists provides an excellent early warning system for the arrival of new species in the UK. While many will be benign and slot into UK ecosystems without causing ecological problems, there are species that have the potential to cause a lot of financial and economic damage. One such current example is a very bright-green beetle, the emerald ash borer, spreading west from Russia. That is definitely one that must be spotted early to allow sanitation strategies to be implemented and avoid disastrous damage to the remaining UK ash trees (*Fraxinus excelsior*), whose population has already been decimated by ash dieback disease (Hill, 2022a).

If we return to the three garden-species tallies we have been discussing, we can ask the question of how the species counts compare with those of other more natural sites such as nature reserves in the UK. In fact, the highest official total species count for any site in the UK is Wicken Fen in Cambridgeshire with 8,674 species. Not only is the site large – well over 1,000 hectares/2,471 acres – but also one where species recording has gone on since 1899. So that puts garden totals of around 2,500 species in a positive perspective. In fact, both Owen's and

Hodge's small gardens are in the top twenty-five of the most species-rich sites in the UK – the others all being very substantially larger and having some kind of nature reserve designation. So, both these gardens are splendid examples of the power of intensive biological recording. Biodiversity deserts? Clearly not.

Chapter 9
The Nature of
Garden Soils

Soil has always been a bit mysterious to an outside observer, but it is clearly essential for most plant life. We have previously mentioned the old adage 'the answer lies in the soil' in relation to growing plants. But what is it that makes a 'good' garden soil? Well, we can think back to one of the pioneers of soil husbandry, who understood how to manage soils to maintain their vitality. Lady Eve Balfour farmed an estate in East Anglia (UK) and wrote a seminal book on the subject, *The Living Soil* (1943), and was also a co-founder of the UK Soil Association. The title of the book emphasized that soil is 'alive' and very much more than inert 'dirt'. To quote from the book: 'Soil is a substance teeming with life. If this life is killed, then the soil literally dies.' Hence, for soil to keep 'living' it needs to be nurtured, in particular through the input of organic material. The core of this life is the soil community of animal and microbial species that are also part of biodiversity in the garden. A recent study has recently come up with the startling figure that soil is home to 59 per cent of the species on Earth. Perhaps a little unconventionally, they included plants as soil species – and it is true that most plant species not only germinate in soil but also have a strong attachment (anchorage, water and mineral nutrition) to it. On that basis they estimated that 85 per cent of plant species, 90 per cent of fungi and 50 per cent of bacteria are soil organisms (Anthony et al., 2023).

So, what can we say about life specifically in garden soil? Not too much actually. While many gardeners may curse moles cruising underground beneath their lawn and periodically throwing up spoil-heaps of soil, the living component of soil is largely hidden. It isn't that scientists haven't studied soil organisms in some detail (there is information about compost invertebrates in Doberski, 2022); it is just that few of these studies have been done in gardens. The results would vary in different parts of the garden – especially in relation to digging/no digging effects. Typically, herbaceous borders and vegetable beds will be dug over to varying degrees each year. This could be seen as 'throwing a spanner in the works' of a stable soil community and stable soil structure of an undisturbed soil. The result might be expected to have an adverse effect on soil invertebrates. Apart from digging, there will also be other forms of soil disturbance, such as weeding, planting, watering, addition of compost, fertilizer and the use of chemical sprays in the garden. The impact of many of these activities on soil organisms (soil biota) in garden settings is largely unknown.

Living organisms in the soil are also likely to be influenced by the nature of the garden soil itself. There is, of course, no such thing as a 'standard' garden soil. Even in one garden the soil will vary from place to place. When a house is constructed and a garden created, the garden soil is likely to reflect the soil type for the area. This is subject to redistribution of soil by the builders, who might, for example, apply a surface layer of 'topsoil' – which can come from elsewhere. The local soil may vary from sandy (the largest mineral particles) to clay (the smallest mineral particles) or something in between. In addition, after years of gardening activity, the soil will have changed. Much of this will be the result of cultivation and the incorporation of organic and other matter into the soil. What starts off as a clay soil may now be a friable loam with a higher content of organic matter. This will be true of the vegetable beds and borders, while the soil under a lawn is likely to be less altered.

All of this means that soil organisms will encounter a range of soil conditions in the garden. This may have consequences for the kinds and abundance of soil biota present. These soil organisms play an essential role in the decomposer food web, collectively processing the organic material entering the soil, promoting the recycling of nutrients, mixing and aerating the soil and thus maintaining soil fertility. This is an example of a critical ecosystem service.

Studies of the effects of soil disturbance on soil invertebrates relate primarily to earthworms, and the issue of digging turns out to defy easy assumptions. We will start by delving a little into the agricultural science literature in which earthworms are considered vital to soil health as they are in gardens. In recent years there has been an active 'to plough or not to plough' debate – that is the question! The argument has increasingly tilted towards no or minimum tillage: eliminating or reducing soil disturbance. What, then, are the consequences of this shift away from agricultural soil cultivation for earthworm populations? The first point to make is that in fields or in gardens we can recognize three groups of earthworm species (Figure 9). Some feed largely on plant litter in the soil surface layers and are known as epigeic worms. Others live in soil at a depth of approximately 0–20cm/0–8in – about the same as plough depth or single digging with a spade. These are the endogeic worms. Finally, we have deep-burrowing anecic worms, which may go down a metre/3 feet or more and drag leaf litter from the surface into their burrows.

Studies on agricultural land (for example, Torppa and Taylor, 2022) have assessed the impact of different tillage regimes on earthworms. Summarizing the results of such investigations, it seems that numerically endogeic worms are least or little affected by tillage. For this group, the potentially adverse effects of soil disturbance may be counterbalanced by the burial of surface plant litter – delivering plant organic matter to the main feeding zone for endogeic worms. The largest adverse effect of

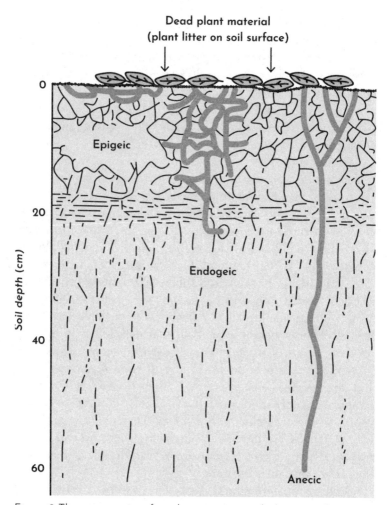

Figure 9 Three categories of earthworm species, which vary in the soil depth in which they feed (adapted from sciencelearn.org.nz).

tillage was on the deep-burrowing anecic worms, presumed to be due to destruction of their deep permanent burrows. One of the consequences of that could be less translocation and mixing of soil from deeper layers – a process known as bioturbation.

Similarly, with fewer deep worm burrows in the soil there may be less gas exchange (oxygen and carbon dioxide) and water seepage deep into the soil. Tillage does, therefore, have some negative effects on earthworm populations, but perhaps not as much as might be expected.

What about gardens and digging? Is there any indication of a similar pattern? A study in Switzerland (Tresch et al., 2018) compared earthworm density across eighty-five gardens, sampling disturbed vegetable/herbaceous beds (with annual plants) and comparing those results for earthworms with areas of uncultivated perennial vegetation (both areas of berries/perennial flowers and grass). Once again, the results were perhaps not as expected. Both earthworm abundance (numbers and biomass) and species richness were highest in the annually cultivated vegetable beds. Species richness and biomass (but not the number of earthworms) were also higher in the uncultivated herbaceous beds than in grass. If anything, cultivation appeared to either enhance the earthworm populations overall or have limited effect. Other data collected during the study suggested that the routine addition of compost to beds by gardeners could have been what contributed to offsetting some of the potentially adverse effects of disturbance. The logic of this seems clear: the incorporation of compost into the upper layers of the soil is likely to provide additional organic matter that will be consumed, especially by endogeic worms.

Once we leave the story of the effects of cultivation on earthworms, the scientific data on other invertebrates get a bit thin. We have already seen with earthworms that tillage or cultivation can result in counter-intuitive changes in earthworm populations. At present it's not clear what would happen to the tiny denizens of cultivated garden soil – for example, the mites (Acari), the springtails (Collembola), the roundworms (Nematoda) and a variety of insects and insect larvae. All contribute to the decomposer community. One thing one can say is that a very compacted soil will not be good for many of these species. If digging or other soil disturbance opens up the soil (increases porosity), that will assist colonization and dispersal by species that, unlike earthworms, cannot easily move through compacted soil. It is also quite likely that the more diverse the above-ground flora, the more diverse will be the range of invertebrates below ground.

In the UK, among larger invertebrates, Smith et al. (2006b) recorded large numbers of woodlice (Isopoda) by sifting through garden-soil surface litter and also in pitfall traps, which sample larger invertebrates moving across the soil surface. Beetles and slugs were also very numerous. These and other groups like springtails (Collembola), mites (Acari) and ants (Hymenoptera) are likely to be ubiquitous in garden soils. These latter three groups were numerically also the most abundant in pitfall-trap studies by McIntyre et al. (2001) in US residential gardens. This was also found in an Australian pitfall-trap study by Norton et al. (2014). Their trial used initially cleared experimental plots which were then set up with four types of ground cover: bare soil, grass, leaf litter and woodchips. All four treatments harboured a range of invertebrates, with grass ahead in terms of invertebrate abundance and diversity. Another study by Byrne and Bruns (2004) sampled mites (Acari) in high-maintenance lawns (mowing, fertilizers and pesticides), low-maintenance lawns (mowing only) and an unmanaged grass field. Curiously, mite numbers were highest in the high-maintenance lawns and least in the unmowed field. The authors cited other studies that showed similar results. One suggested explanation was that the reduction of predatory arthropods in the high-maintenance lawns may have allowed the mites to proliferate. If we think of our ecosystem/food web model of the garden ecosystem, then one can imagine that breaking a strand in the food web – representing a predator–prey interaction – could release mites from predation pressure. Lawns need not be barren of mites!

A study by Salisbury et al. (2020) considered the impact of native versus exotic plant species on surface-active soil invertebrates. They concluded that canopy cover was an important factor in determining abundance of this group of invertebrates.

More cover meant more invertebrates. Although numbers of invertebrates were generally higher on native plant plots, the reverse was true in winter – when exotics provided more cover. So year-round dense vegetation cover was good for many of the surface-active soil invertebrate groups sampled – with some exceptions, such as spiders, which 'preferred' more open vegetation.

Ants have already been mentioned as part of the soil-decomposer community, but they can also play another, less expected role in garden soils. They help distribute plant seeds – a phenomenon known as myrmecochory. They may collect and eat plant seeds directly, but in other cases the plant provides the seed with a small nutritious detachable structure called an elaiosome, which typically contains fats (lipids) and some protein. Only this part is eaten by the ant and the rest of the still viable seed is ejected outside the nest. The elaiosome is a reward (like a sweetie) from the plant to the ant for its contribution to seed dispersal. One example of such a garden 'weed' plant with elaiosomes on the seeds is red dead nettle (*Lamium purpureum*).

So, what conclusions are we left with? In relation to gardening activities, certainly no damning indictment of the ecological horrors of digging. It seems that while some species of earthworms can lose out with less soil stability, others may gain through digging-in of surface organic matter. For other invertebrates the picture is less clear, but addition of organic material to soil and more vegetation cover both appear to have positive effects on soil and soil-surface invertebrates.

Finally, in this chapter, is there a botanical aspect to soil? The obvious one is that soil is the growth medium for garden plants, and growth of those plants will depend on a variety of soil-related factors and overall fertility. But in addition to this, it is worth reminding ourselves that all soils in gardens contain a large store of ungerminated plant seeds – the soil seed bank. When plants shed seeds, some may germinate quickly while others disappear into the soil and may need a period of winter chilling to break dormancy before they can germinate (stratification). But there are many plant species that shed long-lived seeds that can remain buried and dormant in soils for many years. Only when they reach the soil surface – through cultivation – do they germinate if other conditions are favourable. In a tidy garden, many seedlings will be destroyed by hoeing – it is hard to tell 'good' seedlings (ornamentals) from 'bad' seedlings (weeds). A study by Thompson et al. (2005) showed that seeds of perennial native 'weeds' were most abundant in garden soil seed banks. The most abundant alien seeds were from the butterfly bush *Buddleja davidii* – which is not too surprising considering the colonizing tendencies of this plant. It is an interesting experiment to leave an area of garden soil bare to see what comes up. It is one way of diversifying the flora of garden beds and borders.

We have previously alluded to the importance of low soil fertility when trying to create floristically diverse flower meadows. We made the point that nutrient-rich soils tend to become dominated by vigorous grasses rather than flowers. A little more on that topic. We can usefully refer to the classic Park Grass ecological experiment, which is still going strong in 2023 after 167 years at Rothamsted Research in Harpenden, Hertfordshire (UK) (Silvertown et al., 2006). This has been labelled as the world's longest-running ecological experiment. Why is it relevant here? Well, it started off as a grassy field with uniform soil in 1856, which was then subdivided into a large number of smaller plots. A variety of fertilizer treatments were applied to the plots, in particular, nitrogen (N), phosphorus (P) and potassium (K). In addition, some plots were treated with lime, which reduces the soil acidity (pH). Soils not treated with lime become acid (<pH7), those with lime are alkaline (>pH7). The different combination of nutrients and varying soil pH resulted in a mix of vegetation changes, so that adjacent plots could become spectacularly different in appearance and plant composition. There are some plots that look like flower-rich meadows of old. Alongside may be a plot, completely green, with just a few species of grass. This all relates to how they have been treated. There can be fifty to sixty species on the unfertilized plots but only two or three species on some of the plots that have been fertilized. All of this has resulted from a kind of natural 'evolution' of the vegetation over time – some species being better suited than others to particular conditions and some species being excluded by competition by other more vigorous species. Leaving aside the complexity of the data from the various plots, the key messages from this experiment are that to grow a flower-rich meadow you need low-nutrient conditions, particularly with reference to nitrogen (but also phosphorus), and a soil that is not too acidic. Life needs to be tough for all plants on the plot in terms of soil nutrients to prevent 'weaker' species being overgrown and pushed out, but diversity is typically also higher with a benign soil pH (around +/- pH7). The example of UK chalk grasslands makes the point very clearly. They are floristically very rich with soil >pH7 which is dry, low in nutrients and made up mostly of chalk.

Chapter 10
Natural Enemies: A Gardener's Little Helpers

At times a garden can seem to be constantly under attack from 'pests'. The aphids are sucking your roses dry, while the cabbage-white caterpillars are shredding your cabbages! It can all be very frustrating when you have invested time, money and hope into nurturing the plants. Yet not so long ago the UK Royal Horticultural Society suggested in their in-house magazine that the word 'pest' should be banished for good (RHS, 2022b). To convince the sceptical gardener, two arguments were put forward to justify this proposal. The first was that gardeners should recognize that the garden is a complex ecosystem (see Chapter 6) and that we need to be tolerant of all the 'actors' that contribute to its functioning – even if the ecosystem is already heavily modified by garden management. The argument goes that we should promote biodiversity as a good thing per se. But this then leads on to the second point. Although some of our plants may suffer as a consequence of stopping the chemical fight against pests, a more benign approach doesn't mean throwing in the towel completely. Careful and very targeted use of 'eco-friendly' methods of control may be needed at times – but that is a subject for practical books on pest control. Of more interest here is how the ecology of the garden responds when blanket chemical spraying of pests is stopped. Is there a dividend to be earned from a rebound of what are known as 'natural enemies' – a range of invertebrates (also referred to as beneficial insects) and microorganisms that can help to suppress the activity of what we have traditionally referred to as pests? This argument is in the territory of the much-quoted idea of the 'balance of nature'. Although a rather idealized concept that somewhat overstates the case, nevertheless we can go with the idea that organisms eating (or killing) other organisms could have a negative impact on 'pest' species – if we don't kill off the natural enemies. Is that just a nice idea, or is there evidence to back the argument?

The idea of encouraging natural enemies to control pests has, in recent years, gained traction in agriculture in relation to crop pests. As intensive use of pesticides has increasingly eradicated many elements of biodiversity in commercial fields, studies were set up to test whether some of this could be returned by manipulating the field environment. Early trials involved, for example, the creation of grassy strips bisecting arable fields. These strips, known as beetle banks, remained unsprayed and uncultivated and provided a refuge area for a range of predatory insects, especially ground beetles – hence the name (MacLeod et al., 2004). More recent versions of this method have changed to sowing flower-filled rather than grassy strips. Although there is plenty of evidence that such strips enhance populations of potential natural enemies, it has been harder to prove that these lead to a significant reduction of pest damage in the crop. But some studies have demonstrated a clear beneficial effect (for instance, Tschumi et al., 2015).

So, in a garden environment we are looking therefore for three things:

1 ways of increasing abundance and species richness of natural enemies (and, incidentally, also garden biodiversity);
2 evidence that those natural enemies are killing potential pest species;
3 evidence that this effect is sufficient to significantly reduce damage from pest species.

We start by clarifying what is meant by 'natural enemies' in this garden context. Broadly speaking, we are referring to beneficial species that kill pest species in the garden. This could refer to birds eating aphids, for example – they can be encouraged through nesting sites (bird boxes) and bird feeders. In this chapter, we are particularly referring to two groups of invertebrates: insects and spiders. Many insects and all spiders are predatory – capturing and 'eating' (in some cases sucking dry) their prey. However, that is not the only way that invertebrate natural enemies can kill prey. Many wasps (Hymenoptera – bees, wasps and ants) and fly species (Diptera) are classed as parasitoids, rather than the more familiar term parasite. What is the difference? A parasite typically has an interest in keeping its host alive. But insect parasitoids use hosts in a different way, which does eventually kill them. Adult parasitoids lay their eggs on or in another insect of a different species, typically at an immature stage like a larva. The parasitoid eggs can be laid inside the larva of the host species, in the case of parasitoid wasps by using a needle-like ovipositor at the tail end of the abdomen. They may lay more than one egg in (or on, in some cases) the host. The wasp/fly larvae emerging from eggs in/on the host then begin eating their way through the internal host tissues. There is some method in this feeding activity – less 'important' organs/tissues are consumed first to keep the host alive. Only towards the end of the period of parasitoid development does the host die. The parasitoids pupate, undergo metamorphosis and emerge as the next generation of adults (Figure 10). But the story can get more fascinating. There is increasing evidence that parasite and parasitoid species can profoundly alter the behaviour of their hosts to their own advantage. One such example is a *Cotesia* parasitoid wasp parasitizing cabbage-white caterpillars (*Pieris brassicae*). After the *Cotesia* larvae emerge and pupate, the host caterpillar is still alive and spins a silken web over the cluster of wasp pupae to protect them from parasitism by other wasps (hyperparasitism). It also reacts aggressively if disturbed.

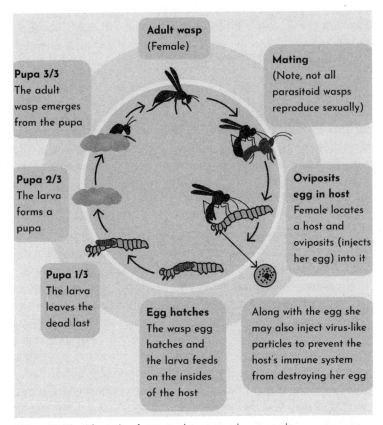

Figure 10 The life cycle of a typical parasitoid wasp with a lepidopteran host (adapted, with permission, from Science Learning Hub – Pokapū Akoranga Pūtaiao, The University of Waikato Te Whare Wānanga o Waikato, www.sciencelearn.org.nz).

The victim (the caterpillar) has been manipulated behaviourally to protect its 'enemy' (the wasps) (Tanaka and Ohsaki, 2006)

Can these types of predatory or parasitoid insects really reduce pest populations in gardens? Well, if you 'seed' a garden with 'test' pest eggs or larvae you can demonstrate that predators (not necessarily just insects) will remove a variable but significant percentage of those prey items – in some cases up to 90 per cent

(Gardiner et al., 2014; Philpott and Bichier, 2017). So, the signs are encouraging, but we need to draw the parasitoids and predators into the garden. It is not too difficult to find studies that show enhanced herbaceous plant cover and floral abundance will enhance predator and parasitoid numbers, as it generally does with other groups of insects. Owen (2010) demonstrated the species richness of parasitoid populations in urban gardens, recording over 500 species of parasitoid ichneumon wasps over a thirty-year period in her urban flower-rich garden in the UK. A study of eighteen urban gardens in California showed that abundance and species richness of parasitoids increased with herbaceous/floral, shrub and tree cover (Burks and Philpott, 2017). But the big question of how this translates to pest control by natural enemies is harder to demonstrate – and evidence is needed to avoid the charge of wishful thinking. Another study by the same Californian research group compared experimental plots with 'messy' versus 'tidy' garden management and their effects on natural-enemy communities and subsequent impacts on pest control (Egerer and Philpott, 2022). The difference between the two types of plots (treatments) related to vegetation cover by 'weeds'. In fact, they found little difference in abundance of natural enemies (or herbivore abundance) between the two systems. There was, however, some compositional difference in species – so the community structure of natural enemies was altered by vegetation composition. They carried out experiments using eggs of a pest species and adult aphids to check the difference in predation levels. Although some differences were observed, the authors concluded that overall the 'tidy' versus 'messy' forms of management lacked a strong consistent effect on pest predation – at least in their trials.

A similar approach compared crop damage in cabbage beds in more or less floral-rich gardens as well as recording herbivore and natural-enemy abundance/richness. Abundance of natural enemies was positively related to floral resources but also to the size of each garden. Overall, there was a low abundance of pests and pest damage across the range of sites and the authors inferred that this was due to healthy populations of natural enemies in all of the gardens (Lowenstein and Minor, 2018). As mentioned in Chapter 3, it looks like mobility of natural-enemy species and other insects can negate small-scale habitat differences within and between gardens.

We have made little specific mention so far of predatory insects. There are a number of different insect taxa in which either larvae or adults are predatory on other insects. Two are particularly worthy of mention, namely hoverflies (Syrphidae) and ladybirds (Coccinellidae). In the case of hoverflies, it is the larvae that are predatory on aphids – which is true of about 40 per cent of UK hoverfly species – while the adults are mostly pollen and nectar feeders. By contrast, with ladybirds both adults and larvae are enthusiastic feeders on aphids. A study by Rocha et al. (2018) studied both hoverfly and ladybird populations in sixty-seven gardens in an urban setting in the city of Reading (UK). Aphids are typically seen as a pest problem to gardeners who can see their plants overwhelmed by large numbers of them, as in the case of rose aphids (*Macrosiphum rosae*) and black-bean aphids (*Aphis fabae*). These herbivorous insects feed by inserting needle-like stylets into the sugar-transporting vein system of the plant (the phloem) and draining the plant of sap. Because aphids need to extract small quantities of useful nutrients (especially nitrogen) from the large volume of phloem sap they take in, excess sugary liquid is voided as sticky honeydew. The Rocha study recorded a total of forty-five aphid species across all of the gardens, although the majority were rare. Broadly speaking, aphid species richness in gardens increased with greater plant-species richness – which might be expected. The more host plants species, the more varied the aphid species. The pollen- and nectar-feeding adult hoverflies were more abundant with increased plant-species richness, which was also seen with ladybirds, albeit the trend was less clear-cut. Although the data were quite scattered, higher ladybird abundance was generally associated with aggregations of aphids. So, there can be plenty of ladybirds and hoverflies in gardens with aphid-infested plants. But do these two groups of beneficial insects actually contribute to reducing aphid numbers? There was no direct answer in this study, but the answer is clearly 'yes' in the sense that hoverfly larvae and ladybirds can often be seen feeding on aphids. But typically, the aphids have a good head start in building up numbers before the hoverflies and ladybirds arrive. Aphids have a complicated life cycle, but during population growth they can reproduce parthenogenetically (without mating). So, while plugged into their phloem food source, the female aphids generate a continuous production line of young aphids. Consequently, their populations can increase very quickly; and typically the arrival of hoverflies and ladybirds can stall some of this increase, but not halt it. So, a fastidious gardener may still need to revert eco-friendly soap solution with which to spray aphids – after first checking for larvae of ladybirds and hoverflies.

Taking these studies as a whole, direct evidence of natural pest control by promoting natural enemies in gardens is still rather thin. This might be a rather disappointing result for advocates of more naturalistic gardening for wildlife, but it demonstrates the relative dearth of research in this area. There is a need to answer the question of whether stopping use of pesticides, increasing floral richness and reducing management intensity increases effective 'natural' pest control. With little published evidence from appropriate surveys and experimentation in real gardens, the jury is still out. For gardeners it is a case of experimenting – doing the right things to promote natural enemies. In combination with other non-chemical methods touted for pest control and a degree of tolerance of pests, it may all come together 'naturally'.

Chapter 11
Garden Plants, Herbivores, Pollinators and Pollination

We have already seen how much of the hidden diversity of gardens resides in invertebrate species, with some very impressive totals that naturalists have recorded in gardens, especially insects. Of course, the fact that an insect is captured in a garden doesn't necessarily mean that it is 'at home' in a particular garden setting. With the range of flying insects seen in a garden, some species will turn out to be 'tourists' – just passing through the garden. Some may test the 'facilities' and decide to stay. A herbivorous insect landing on a plant can often 'taste' a plant's suitability for feeding and breeding. You might expect the taste sense organs (gustatory receptors – tiny hair-like structures called sensillae) to be around the mouth, but they can be elsewhere. Flies and lepidopterans (moths and butterflies), for example, have taste receptors on their leg tarsi – which amounts to having them on their 'feet'.

As far as many herbivorous insects are concerned, the characteristics of plants in the garden will dictate their suitability for residency. We need to remind ourselves that although much of the world is very green thanks to plants, those plants have had to survive the ravages of hungry insects (and other herbivores) to still be with us. So, an evolutionary 'arms race' between plants and herbivores has ensured that generalist herbivores must be picky about which plant they choose, because many plants have physical or chemical defences that they need to avoid. The chemical defences come from a group of secondary metabolites (referred to earlier in Chapter 5), of which thousands are known, synthesized by a range of different plant species. Many of these metabolites have toxic or repellent properties. Some plants are loaded with many more chemical defences than others, so there will be plants with more limited chemical defences. Specialist herbivores have adapted to feeding on a group of related (or even a single) plant species – and remain unaffected by defensive chemicals that are toxic to other herbivores. One example

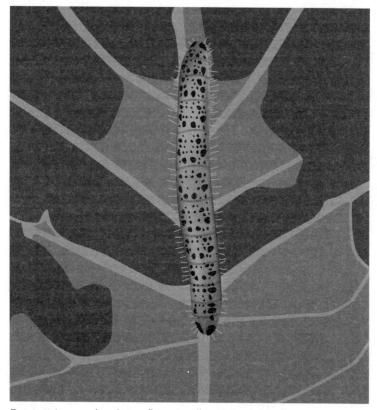

Figure 11 Large white butterfly caterpillar (*Pieris brassicae*) are adapted to feeding on plants of the Brassicaceae family, which contain defensive glucosinolate chemicals.

is herbivorous insects feeding on plants of the cabbage family (Brassicaceae/Cruciferae). Plants in this family contain a group of defensive chemicals called glucosinolates. Not only do insect 'pests' such as cabbage-white butterfly (*Pieris rapae*) and cabbage aphid (*Brevicoryne brassicae*) cope with glucosinolates when they feed (Figure 11), but the cabbage aphids use them (sequester) for their own defence. This process of adaptation to plant chemical defences helps to explain why many herbivores restrict feeding to a single plant family or even one genus or species.

So, the range of plants in the garden will be an important factor (albeit not the only one – a number of other conditions need to be right) in affecting which herbivore species can complete much of their life cycle in the garden. But in gardens the situation is normally complicated by the presence of many non-native flower species originating from many different parts of the world. The majority are from Europe, but species from Asia and North America are also strongly represented. Southern-hemisphere contributions come from New Zealand and South Africa (Thompson et al., 2003). For example, in Owen's garden (Owen, 2010), about 60 per cent were alien plant species and about 40 per cent native. The results from the BUGS study of 61 gardens (Smith et al., 2006c) appeared to shift the balance of plants more towards alien than native plants in gardens – 30 per cent native and 70 per cent alien. But raw numbers of species for the whole study can hide a different truth. Because many of the aliens were recorded only in single gardens (79 per cent), in a typical garden on average 45 per cent of the plant species were native. Nevertheless, non-natives still predominated. This is perhaps not surprising since 81,000 plant species and varieties are available for gardeners to buy from plant nurseries and garden centres in the UK (RHS, 2022a). This compares with around 1,600 UK native plant species.

Because alien plants are the majority in gardens, there has been much debate about the usefulness of these alien plant species as a resource for garden wildlife. This is one of the questions we will focus on in this chapter.

Enthusiastic gardeners in the UK and beyond plant many different flowers, mostly alien, grown from seed or purchased as mature plants. But in recent years much thought has been given to making gardens 'insect-friendly' – especially for pollinating insects on the lookout for pollen and nectar food sources. As they feed, they perform their pollination function – an essential ecosystem service (see Chapter 7). Many crops (including garden crops) rely on the services of pollinators extensively. In temperate regions, around 78 per cent of plant species can be animal pollinated, the vast majority of which will be by insects (Ollerton et al., 2011). Insects are typically thought of as providing a service to plants for cross-pollination, but they can also contribute to self-pollination. Perhaps surprisingly, overall around 10–15 per cent of plants predominantly self-pollinate (Wright et al., 2013).

What can we say about the value to pollinators of sown or purchased alien plants? One study suggests that for the ecologically minded gardener, there could be a slight feeling of guilt as they buy an exotic plant from a garden centre. The title of the paper 'Most ornamental plants on sale in garden centres are unattractive to flower-visiting insects' (Garbuzov et al., 2017) seems to say it all. But this general message masks the presence of a significant number of pollinator-attracting alien species. How can one tell which plants fit that category? That is much easier now with the use of a variety of logos on plants said to be attractive to pollinators (for example, the 'RHS Perfect for Pollinators' is one among several different labelling schemes). And yes, in the Garbuzov study those pollinator-friendly plants were typically 4.2 times more attractive (median values) to pollinators than other alien species – which was positive. Less good news was that not all plants with such labels proved attractive to pollinators, on the basis of data in this study. This suggests that there may be some room for further testing and refinement of the labelling systems.

The UK Royal Horticultural Society has a keen interest

in the value of alien garden plants for pollinators. The welfare of pollinating insects is one aspect of garden ecology that has received a lot of publicity in recent years. While the initial emphasis was on honeybees and bumblebees, more recently attention has turned to other pollinator groups such as solitary bees and hoverflies. The RHS initiated a study that reported on the attractiveness of non-native plants to pollinators and the attractiveness of non-native plants to other types of plant-associated insects. The results appeared in two scientific papers (Salisbury et al., 2015; 2017). The researchers tested the attractiveness to insects of plots containing three different plant selections (three treatments): native, near-native (from the northern hemisphere), and exotic (only from the southern hemisphere). What did they find? Starting with the response of pollinating insects, as might be expected, on a series of experimental plots, more flowers equated with more pollinator visits – irrespective of the geographical origin of the plants. Comparing the abundance of insects across the treatments, the native and near-native plants were similar – but the exotic plots were about 40 per cent lower. However, there were differences between different groups of pollinators – there was no 'one size fits all' result. Honeybees 'preferred' the near-native treatments. Hoverflies were most abundant on the native plots. The 24 species of bumblebees in the UK varied in their preferences for flowers, which can be partly related to differences in tongue length between species, which can vary by a factor of around two. So, short-tongued bumblebees made considerably fewer visits to the exotic plots. One explanation for this can be seen from a study by Tew et al. (2022) of seasonal nectar production by common UK garden flowers. They found, for example, that the exotic hybrid *Fuschia magellanica* plants, commonly grown in UK gardens, are copious producers of nectar. But they are inaccessible to short-tongued bumblebees, which avoided the flowers. Of course, there are honeybees and short-tongued bumblebees that are a little sneaky and have learned to chew holes at the base of tubular flowers to steal the nectar! It must be remembered that many insects (such as honeybees and species of bumblebee) have a distribution range that extends beyond the UK in mainland Europe. So those same honeybee and bumblebee species (as in the UK) will be foraging among near-native plant species that are native in mainland Europe and so presumably can cope with such near-natives grown in the UK.

As noted, the RHS study found that the exotic plots fared less well overall in terms of attracting pollinators. But that isn't quite the full story. The exotics often flowered longer and did help to extend the season for insect foraging into late summer (September), when flowering on the other two treatments was in decline. I have seen the same in Cambridge Botanic Garden, where plots sown with native seed mixtures typically ceased flowering earlier than flowers from seed mixtures with a range of alien species. So, extending the flowering season is an important 'plus' for growing non-native plant species in gardens – as long as they are species that are still attractive to pollinators.

A similar study in Maryland (USA) (Seitz et al. 2020) on native versus non-native plants for pollinators showed mostly similar results. They found that a non-native seed mix of pollinator-friendly plants were well accepted by a wide range of wild bees. In early and late season, they could even attract more individuals and species than native plants could, with little difference in the summer. However, the visiting bee community on native versus non-native plots was different. More specialized bee species were recorded on native plots; that is to say, species adapted to particular flower types. This would be expected on the basis of coevolution of flowers and pollinators.

Unlike the studies already quoted, Rollings and Goulson (2019) found no difference in numbers of insects attracted to native or non-native garden plants (25 native out of a total of 111 plant types), although flowering native plants attracted a higher diversity of insects. They did find differences in which kinds of insect preferred which plants, even between closely related species of insect. Why the difference between studies? We need to be sure that we are considering like-with-like when comparing the results of different studies. In this case, individual plant species were tested separately for their attraction to pollinators. Other studies, such as that by the RHS, compared overall attractiveness to pollinators of different mixes of plant species grown in 'meadows'. So, the Rollings and Goulson study didn't test the combined effect of a range of flowering 'meadow' species attracting pollinators over a growing season. But it did tell you about the attractiveness to pollinators of individual plant species. We have already noted that extending the flowering season in gardens through growing non-natives is likely to promote pollinator abundance and diversity.

Of the plants Rollings and Goulson tested, those with the highest total number of visits were lesser calamint (*Calamintha (Clinopodium) nepeta* – native), autumn sneezewort (*Helenium autumnale* – non-native) and a hybrid cranesbill (*Geranium* 'Rozanne' – non-native).

Whichever flowers are grown in the garden, the more diverse the range of flowering species, the more even is the nectar supply from early spring to autumn. And the more 'open' the flower structure, the more accessible are the flowers to a wider range of pollinators – for example, bumblebees, hoverflies and solitary bees. Tew et al. (2022) suggested a list of flowers (based on UK gardens) to provide nectar over the full growing season (Table 4).

As well as filling the growing season with flowers, choice of flower mix might also be influenced by knowing which species produce abundant nectar and/or pollen. A very useful study published by Hicks et al. (2016) does just that. It provides very detailed data on sixty different flowering-plant species – classed as annuals, perennials and weeds. The annuals and perennials were species sold in commercial seed mixes for planting up flower-rich meadows. The list included both UK native and non-native species. All had their pollen and nectar production

Seasonal period	Recommended plants	Native status	Flower structure
Early spring (March)	*Helleborus* spp.	N or A	S
	Pieris spp.	A	S
	Pulmonaria spp.	N or A	S
	Salix spp. (willow)	N or A	G
	Skimmia japonica	A	G
Mid to late spring (April-May)	*Aquilegia vulgaris*	N	S
	Ceanothus spp.	A	G
	Malus spp. (apple)	N or A	G
	Prunus avium (cherry)	N	G
	Ribes spp. (currants)	N or A	G or S
Early to mid-summer (June-July)	*Campanula* spp. (bellflower)	N or A	G
	Geranium spp. (cranesbill)	N or A	G or S
	Lavandula spp.	A	S
	Lonicera periclymenum (honeysuckle)	N	S
	Pyracantha coccinea (firethorn)	A	G
Late summer to autumn (August-October)	*Echinacea purpurea* (coneflower)	A	G
	Hedera helix (ivy)	N	G
	Origanum vulgare	N	G
	Sedum spp.	N or A	G
	Verbena bonariensis	A	S

Table 4 Recommended plants for different seasons in UK gardens. Species labels indicate native (N), non-native (A) and generalized (G) or specialized (S) flower structure (Tew et al., 2022).

carefully measured – which generated some interesting results. The 'top three performers' in each category are shown in Table 5. Several weed species produced particularly large quantities of nectar per 'flower' (or floral unit), including ragwort, two thistles and dandelion. The use of the term 'floral unit' referred to the fact that in some species the 'flower' is actually an aggregate of very small flowers; this is true of the daisy family (Asteraceae/Compositae) in particular. You can see those tiny florets with a lens if you tease apart a daisy or dandelion flower. Two native perennials, a hawkbit and common knapweed, also produced attractive quantities of nectar, while annual flowering plants were relatively poor in nectar production compared to the best 'weeds' and some perennials. Differences in pollen production were generally less marked, but the top five included three annuals – common poppy (archaeophyte), California poppy (non-native) and corn marigold (archaeophyte) – and two perennial natives – musk mallow and common knapweed. The only weed that stood out a little from the other weeds in higher pollen production was dandelion – which, incidentally, also produced a lot of nectar. These data show that some species of flowers that produce a lot of nectar are not so good on pollen production and vice versa. But one rule of thumb, evident from Table 5, is that plants from the daisy family (Asteraceae/Compositae) are typically good all round. They represented fourteen out of the eighteen species listed.

Another study from the USA (Majewska and Altizer, 2020) collated a wide range of scientific publications on the subject of insect pollinators and gardens and highlighted a similar set of characteristics to encourage pollinators. The main factors were high plant-species diversity, woody vegetation and garden size – but with the addition of sun exposure and closeness to natural sites. Sun exposure encourages early morning 'basking' by flying insects and higher temperatures for flight – hence more pollinator activity. Closeness to natural sites can potentially broaden the influx of species into the garden. Sunlight

Table 5 Three most productive plant species in each category:

	Nectar			Pollen		
	Common name	Latin name	Nectar µg/day	Common name	Latin name	Pollen µl/day
Annuals	Cornflower	*Centaurea cyanus*	896	Common poppy	*Papaver rhoeas*	6.0
	Cosmos	*Cosmos pinnatus*	701	California poppy	*Eschscholzia californica*	2.4
	Pot marigold	*Calendula officinalis*	470	Pot marigold	*Calendula officinalis*	1.8
Perennials	Rough hawkbit	*Leontodon hispidus*	1827	Musk mallow	*Malva moschata*	2.3
	Common knapweed	*Centaurea nigra*	1474	Common knapweed	*Centaurea nigra*	2.1
	Vipers bugloss	*Echium vulgare*	688	Ox-eye daisy	*Leucanthemum vulgare*	1.1
'Weeds'	Common ragwort	*Senecio jacobaea*	2921	Dandelion	*Taraxacum agg.*	1.3
	Creeping thistle	*Cirsium arvense*	2607	Rosebay willowherb	*Chamaenerion angustifolium*	0.7
	Spear thistle	*Cirsium vulgare*	2323	Corn marigold	*Glebionis segetum*	0.6

Table 5 A trial of sixty flowering plant species from seed mixes to measure their nectar and pollen production. The top three most productive species are listed in the table under the headings Annuals, Perennials and 'Weeds' (Hicks et al., 2016).

has also been cited in a study on bee- and butterfly-species richness in an inner-city area (New York, USA) (Matteson and Langellotto, 2010).

The RHS study identified several trends concerning 'plant-associated invertebrates'. In general, the native plants had the higher abundance of invertebrates (about three times higher than exotics) but for some groups there was little difference compared with the near-natives. However, an important additional factor was canopy cover: dense vegetation yielded more insects. Although exotics did less well overall, they still supported a suite of species and hence were a valid addition to the floral mix in the garden. One alien plant that makes this point very well is the butterfly bush (*Buddleja davidii*), originally from China. Not only is it a great source of nectar for many butterfly species but Owen (2010) also found that it was the plant with the widest range of species of herbivorous moth larvae in her garden. It was fed on by the larvae of nineteen moth species, despite being an alien species.

However, the picture may not always be so rosy for herbivores feeding on non-native plants. A study from the US found that local moth species did not fare well on non-native plant species when tested in feeding experiments with larvae of four species. The results were variable, depending on the moth species, but there was either no survival of larvae or a reduced rate of development (Tallamy et al., 2010). How these experiments relate to the garden situation is not clear, but they suggest that there are limits to the flexibility of herbivores feeding on non-native host species. In practice, some alien plants have a lot more herbivores associated with them than do others. Being related taxonomically to a local species makes a difference, and so species within the same plant family are most likely to share native herbivorous species.

All of these results generally fit in with those expected from the earlier BUGS study (introduced in Chapter 2). Recommendations resulting from this study for improving gardens for wildlife

(Gaston et al., 2007) covered some similar points. The three-dimensional complexity of vegetation was deemed important. Although the reference was to trees, shrubs and hedges, it echoes the value of canopy cover referred to in the RHS study. In relation to the floral mix, the Gaston study noted that natives are excellent for plant-feeding taxa, but other species such as predators, parasitoids and pollinators are not so dependent on the origin of particular plants. Even among herbivores there are generalists that can feed on exotic plants. What many insects do need is a pollen and nectar resource. So, to quote from Gaston et al.: 'Gardens with few native plant species can be just as rich in invertebrates as those with many native plants.'

Now we return to the value of trees and shrubs mentioned in the BUGS study. Addition of these to a garden has previously been noted as of great value for biodiversity (see Chapter 3), but inevitably mostly a feature of larger gardens. Trees can provide resources for a range of invertebrate species that don't make a home in herbaceous vegetation – for example, as a feeding site for a variety of herbivorous caterpillars of moth and butterfly species. You can imagine that for each unit area of ground, trees produce far more biomass of vegetation than herbaceous species. That helps to explain why so many UK bird species – for example, blue tit (*Cyanistes caeruleus*) and great tit (*Parus major*) – rely on trees to provide a feast of caterpillars at nesting time. It's harder to think of trees as a source of nectar and pollen because many are wind pollinated – oak (*Quercus robur*) and birch (*Betula pendula*), for instance – and so don't rely on insects. That said, UK species like sycamore (*Acer pseudoplatanus*), rowan (*Sorbus aucuparia*), field maple (*Acer campestre*), lime (*Tilia* spp.) and wild cherry (*Prunus avium*) are insect pollinated. And, although it is more often seen as a low-growing shrub or as a hedge species, hawthorn (*Crataegus monogyna*) is very common in the UK and attracts throngs of insects to its flowers in May.

What about native and non-native trees and wildlife in a broader context? Are non-native species of value in the garden for promoting general biodiversity? A review of tree species and their associated wildlife was published by Alexander et al. (2006). This ranged across many different kinds of organisms, including fungi and lichens as well as invertebrates. Most of the species covered were native, but also included were about ten non-native tree species commonly found in parks and gardens. In their concluding remarks the authors stated that 'most [tree] species are of significant value to wildlife, irrespective of whether or not they are native to a particular area of Britain.' They did, however, note that the two least 'useful' species were non-native walnut (*Juglans regia*) and turkey oak (*Quercus cerris*). Of course, there are many more non-native tree and shrub species grown in UK gardens. Some may belong to the same genus (congeneric) as species already present in the UK – for example, maples (*Acer*). Other more exotic species may have no clear taxonomic link with any UK plant species and so are less likely to suit native herbivores.

The overall message of this chapter is clear. There will be rewilding enthusiasts who revert their gardens to a native-only status, but UK gardens will continue to showcase a mix of plant species from near and more distant temperate geographical regions. Some will attract more insect species and some fewer – particularly if they are categorized as non-native. However, most plant species have something to offer visiting insects, so a purist approach to grow only 'bee-friendly' or 'wildlife-friendly' plants is not advocated. On the other hand, growing a garden full of exotic subtropical plants will do rather less for garden wildlife. A good mix of all kinds of native and non-native herbaceous and woody species will both provide for wildlife (most likely by creating a longer flowering season) and provide aesthetic and ecological interest.

Chapter 12
Succession and Rewilding

There are several reasons why a gardener may wish part or all of their garden to 'go wild'. Maybe the garden is too large or too much work, or there's the belief that wilding a garden is 'good for nature'. What we know is that the rewilding (or wilding) of areas of farmland or other countryside is much in vogue in the UK and elsewhere for the promotion of 'naturalness'. This notion is then linked to a variety of benefits, including enhanced biodiversity. There are a number of such projects in the UK in different types of habitat. One in particular has become very high profile. A farm in Sussex, England (Knepp Estate) is being rewilded on a large scale (951 hectares/2,350 acres), at least by British standards. This is not simply a case of letting nature do its own thing – the rewilding here is choreographed to achieve specific objectives (Tree, 2018).

What exactly does rewilding (wilding) mean? This is tricky to answer: there are many different versions of the concept. At its most straightforward, one might think of it as allowing managed land to revert to a 'wild state'. In the UK, there are some examples of where this has been done. One such location is at Rothamsted Research in the town of Harpenden, which has been known for its pioneering work in agricultural research since the nineteenth century. That apart, we can focus on two small strips of agricultural land (Broadbalk Wilderness and Geescroft Wilderness) that were abandoned in the 1880s and allowed to revert to nature. The process of reversion, involving gradual change in vegetation and associated animal fauna, is called ecological succession. A mix of plants grows up on the vacated land, initially from the seed bank (see Chapter 9). That is the reservoir of ungerminated seeds already present in the soil. These may be remnants of a previous crop as well as a variety of herbaceous species – especially those normally labelled as weeds. Trees and shrubs also begin to appear from buried seed or are brought in by birds or mammals. Those newly growing trees and shrubs would not have survived regular ploughing in previous years when the land was cultivated. Over the years, the species present on the land continue to change in tandem with changes in the soil (for example, organic matter, nutrients, pH) and the environment (shading as trees and shrubs start to dominate, for instance). Over much of the UK, the 'end' of the ecological succession process would typically be high-canopy woodland. This is what has happened on these two fragments of agricultural land. The resulting woodland has tree species local to the area, such as field maple (*Acer campestre*), hawthorn (*Crataegus monogyna*), oak (*Quercus* spp.) and ash (*Fraxinus excelsior*). In much the same way, any sizeable area of garden left entirely to its own devices would most likely reach the same end point: a woodland, which is then referred to as the climax vegetation. But it might take a lifetime!

Although there is a degree of ecological reverence for habitats that haven't been messed with by man – for example, tropical rainforests or 'virgin' forests in mainland Europe (such as Białowieża Forest in Poland), in reality most 'natural' areas remaining in the UK and mainland Europe have been moulded to a lesser or greater extent by human activities. Flower-rich meadows are the product of a rigid management regime of hay cutting and grazing. They never reach an ecological climax but are instead kept perpetually 'young' in succession terms – this is known as a deflected succession. And so it is with rewilding. In most cases it would not be considered desirable to have a reversion to an uncontrolled process of succession – at least not beyond a certain point. There is often a wish to create a mixed set of habitats on the land, each requiring the imposition of a particular management regime.

From an academic standpoint, there are two rather different concepts of rewilding. At one end of the spectrum, rewilding links with the concept of 'wilderness'. This is rewilding applied to extensive land areas where people have or have had little impact and where land can be maintained or reverted to a more or less 'natural' state. The areas of land are large enough to have their own dynamic and be a self-sustaining ecosystem, with predators to keep prey populations in check. This type of approach is essentially impossible to apply in populated areas of Europe, North America and many other parts of the world. The alternative view of rewilding takes a more nuanced approach. This accepts that rewilding must integrate the needs of people with nature and move towards a managed form of rewilding but with the essential aim of giving nature more autonomy (Schulte to Bühne et al., 2022). An example of this might be taking away some fences and other obstacles to give grazing herbivores free range over more extensive areas of land, with less human 'interference'. Ecological changes on the land are then mostly driven by natural regeneration of flora from seed or vegetative reproduction (such as tree-root suckers) and the activities of their associated fauna. The fauna may arrive naturally, but in the case of large herbivores, these are often introduced into the area. However, because the process remains managed, what are thought of as the wrong kinds of changes may still be redirected. For example, there may be too many herbivores and overgraz-ing, so stocking rates need to be controlled. The end result, with this approach, is at best 'semi-natural' rather than nature doing its own thing – but is still definitely wilder than before. And to be thought of as rewilding, it needs to be less controlled than conventional conservation management where the outcomes are more planned. In some cases, the distinction between the two may be a subtle one. There is full discussion of these issues in Fuller and Gilroy (2021).

So, what about gardens? The term 'wilding' is more appropriate than rewilding because in most cases we are not attempting to return the garden to some kind of earlier condition. The wilding is typically on a small scale and with limited ambition. The scope will depend on the size of the garden, but a part of the garden might be left to revert to a 'natural' vegetation – encouraging the appearance of wild, 'weedy' or self-sown flowers, grasses and a variety of associated species. This may prove more or less 'successful' from a gardening perspective, depending on initial expectations, but may not conform to a picture-postcard flower-rich meadow. That doesn't mean that even a slightly untidy patch of mostly grasses is not a biodiversity gain: a number of grass-eating moth caterpillars could be very grateful, while other invertebrates may appreciate the dense cover. But most gardeners will still want to tidy up by mowing at the end of the growing season after plants have shed their seeds. If the grass cuttings are removed, this can lead to gradual depletion of soil nutrient levels, which in time might result in more floral-species richness. Mowing will also prevent tree and shrub colonization. The area will then remain as a grass/flower mix from year to year – combining the attributes of a little 'wildness' with the practical requirements of garden management.

One alternative is to sow a patch of 'meadow' on a small scale (we discussed meadows in Chapter 3). The issue of soil fertility was mentioned earlier as an important factor in determining the success of growing wild flowers. To stick with the ethos of wilding, there is one more issue to consider if the intention is to sow a native-plant meadow: the seed mix. There are increasing concerns about the origin of those seeds – known as the seed provenance. Plants grown in a different region or country may be ecologically different (this can be referred to as having a different ecotype), so, for meadow creation beyond garden scale, locally harvested seed should be the main option.

Such larger-scale habitat creation can be done by seed collection from local nature reserves, but this is more difficult to apply in a small-scale garden setting. At present, labelling of such meadow seed mixes for garden use may provide little information about the origin of the seeds.

We can see that the idea of wilding a typical garden is going to be some way from being 'natural' or a wilderness. But the idea has caught the public imagination and generated a lot of controversy, about both the biodiversity value of such wilding and how it relates to the idea of gardening and gardens. Maybe we can see why if we consider a dictionary definition of the word 'garden': 'a plot of ground – usually near a house – on which plants (flowers, vegetables, fruits or herbs) are cultivated'. The term 'cultivated' implies a high degree of management effort to achieve a planned garden with different areas for the different types of plant. In many cases this would also include an area of grass represented by a highly manged lawn. In addition, many gardeners aim to achieve what they consider an aesthetically pleasing outcome in the garden. Applying this definition of a garden, a number of gardening celebrities have questioned whether wilding an area can ever result in a 'garden'. The truth is that if someone has a piece of land alongside their house and they choose to 'wild' it, then for them it is still a garden. Maybe the definition of a garden needs change. Having said that, there remains the question of whether wilding a small garden is good for wildlife. All the scientific background in this book suggests that a garden that has been 'let go' to become more wild will undoubtably be attractive to a range of species, but a rather uniform wild garden may not be particularly species-rich. That is a generalization, the truth of which will vary depending on the nature of particular gardens. However, one well-known gardener has commented that such an approach would be 'catastrophic' for wildlife. We noted earlier that extending the flowering season by planting exotic plants rather than relying just on native species is a positive for insects reliant on nectar and pollen. This and other evidence suggests that a garden with mixed planting as well as some wild areas is likely to be attractive to the widest range of species. So, if you set about recording the species richness of your patch of garden, diverse planting as well as some wilding would likely give you the highest tally. Ultimately, a degree of wilding in a garden is not a disaster for wildlife, but creating a range of more formal plantings with both native and non-native species alongside less-managed wild areas is the optimum mix. There have always been different ways to garden, and wilding is an addition to the smorgasbord of gardening options.

Chapter 13
A Resumé: Discovering and Promoting Biodiversity in Gardens

We have discussed a variety of topics that touch on what promotes biodiversity in gardens. This chapter provides a 'final word' recap of what to consider if you feel your garden is not up to scratch in doing its bit for biodiversity.

Maybe the first point to note is that you may be underselling your efforts. As is clear from Chapter 8, there may be a lot more hidden biodiversity in the garden than you imagine. You may wish to take on the challenge of finding out more about what is in your garden. The plants are there in front of you – so you just need a good garden flower and field guide to start you off. The invertebrates do need more specialist equipment, such as a moth light trap or a Malaise trap to catch a range of flying insects. These are relatively serious investments but a sweep net for swishing through the vegetation or a butterfly net will be more affordable. Pitfall traps dug into the soil can be improvised from discarded plastic cups and used for insects that live on the soil surface, such as ground beetles (Carabidae). Water-filled pan traps can be used to catch a variety of flying insects on the basis of their attraction to particular pan-trap colours. A brief summary of these methods and equipment is given in the Appendix.

As well as field guides to help with identification, there are published identification keys, specialist websites and Facebook groups where advice and help can be sought. In recent years, a number of phone apps have appeared that rely on matching your photograph of a plant or 'bug' to a large database of species photos. Having been initially a little sceptical about their effectiveness, I suggest they can give a surprisingly accurate identification – at times. But the apps can get things wrong, so they should be treated as an aid to identification but not necessarily a final arbiter of species name. If you have a name and a photograph (many mobile phones can take good 'macro' photos), it is a good idea to submit the record to a database such as the 'i-record', run by the UK National Biodiversity Network (NBN). An app with a more global reach is 'iNaturalist', based in the USA. In addition, it is worth exploring possible involvement in citizen-science (also referred to as community-science) projects. Typically focussed on one taxonomic group (for example, ladybirds or bumblebees), these projects make the task of identifying species in your garden more manageable than a broad-brush approach. It's a good starting point. And then the more you know, the more you may wish to know.

But what is it about the way that gardens are managed for biodiversity that can be improved? Aronson et al. (2017) list the following negative aspects of garden management:

1   maintenance of a grass lawn;
2   removal of habitat, including pruning and leaf litter removal;
3   simplification of habitat structure;
4   pesticide and herbicide applications.

We can link these statements to research cited in the preceding chapters:

Lawns: Even a short lawn will have some floral interest if not cut to 'bowling green' length and can provide flowers (such as white clover) for visiting bees and other insects. In Chapter 3 we discussed the concept of tapestry lawns. These replace grass with other mowing-tolerant flowering plants and create a tidy carpet of flowers that is also resource-rich for insects, especially pollinators. Nevertheless, areas of more diverse and longer grass are also essential for many species, both for feeding and habitat (cover). It will provide a habitat for caterpillars of many moth species, such as the micro-moth family Crambidae. The habitat function of longer (untidy!) grassy areas will harbour insects such as ground beetles (Carabidae). And, of course, uncut areas of grass may include a number of native flowering-plant species that appear in the grassy mix. Some flower/weed seed can survive in soils for decades and germinate and grow after intensive mowing stops.

Removal of habitat: This is part of the tidiness culture that pervades many gardens. There is a balance to be struck. Leaves that fall on flower and vegetable beds will be incorporated into the soil by earthworms. They can also provide an overwintering location for 7-spot ladybirds. If piled into a compost heap (mixed with grass cuttings), they will not only provide compost to enhance the organic content of soil but will also contribute to a compost ecosystem that will have its own suite of species likely to differ from other parts of the garden (Doberski, 2022). One of the key biodiversity gains in a garden is to leave fallen wood/branches/dead standing trees that are allowed to decay. This provides a substrate for a large saproxylic community – species of wood-decaying fungi – as well as many different types of invertebrates that feed directly on the dead wood or consume fungi or wood colonized by fungi. If there is a need for tidiness, then dead wood can be stacked in the wood pile – providing both a feeding site for some species and shelter for others.

Weed removal also equates with removal of habitat. Retaining a patch of nettles or other 'weeds' provides host plants for a variety of insect species to feed and breed. However, it is worth bearing in mind that insects can be fickle creatures and there is no guarantee that they will appreciate your largesse. Your carefully preserved patch of nettles may remain uninhabited by larvae of 'showy' butterflies, such as red admirals (*Vanessa atalanta*), painted ladies (*Vanessa cardui*) and commas (*Polygonia c-album*). But there will always be some insect life enjoying your nettles. A nice guide to identifying these insects is available (Davis, 1991).

There are ways in which gardeners can create a habitat for particular species, such as providing bug hotels. Mine has worked well, attracting good numbers of mason bees (*Osmia bicornis*) to lay eggs inside hollow bamboo canes. Unfortunately, the local birds have now learned the trick of pecking some of them out of the canes. So, I have inadvertently provided them with a takeaway venue!

Simplification of habitat structure: There are a variety of aspects to this. If you remind yourself what semi-natural vegetation looks like – let's say in a nature reserve – every square metre will be stuffed full of plants. Many of these will be grasses, but they will also be interspersed with 'wild' flowering plants. By contrast, a herbaceous border will be carefully weeded to ensure that there is only bare earth and the specimen herbaceous plants. With the removal of any competing plants, the specimen plants grow large and suitably showy.

One of the key findings of the Salisbury et al. (2017) study (Chapter 11) was the role of canopy cover in gardens for enhancing invertebrate abundance. By definition, you are not going to get a dense and complex canopy in a tidy herbaceous border. On the other hand, this can be achieved by wilding part of the garden (Chapter 12) retaining some weeds or seeding parts of the garden with a mini-meadow seed mix (Chapter 3). These approaches can increase the richness of insect species in the garden and, in the case of mini-meadows, encourage both pollinators and natural enemies. Another aspect of habitat structure is the addition of trees and shrubs to a garden. This enhances the canopy complexity and provides a habitat for a new suite of invertebrate species, especially moth caterpillars. It was not surprising that the BUGS study (Chapter 3) found trees were more common in larger gardens, which have more space. Although larger gardens were not more attractive to insects per se, the presence of trees and hence a more complex habitat structure should enhance species richness.

Pesticide and herbicide application: All biocides are potentially toxic to varying degrees to a variety of insects and other invertebrates and to plants. There is a clear suggestion (though difficult to prove categorically) that the significant decline in insect abundance noted for some insect groups and species in Europe and North America is linked to exposure to biocidal agrochemicals. One recent study surveyed 615 UK gardens and found that 32 per cent of gardeners still employed biocides to control 'pests' and 'weeds'. Use of pesticides (especially the herbicide glyphosate and the molluscicide metaladehyde) appeared to correlate with reductions in numbers of house sparrows in gardens (*Passer domesticus*) (Tassin de Montaigu and Goulson, 2023). For ecologically minded gardeners, stopping the use of synthetic biocide in gardens seems a prudent step in view of the doubts surrounding their large-scale use in agriculture. Where necessary, there are often acceptable alternatives approved for organic horticulture. Again, all this links in with the idea of seeing urban gardens as refuge areas for encouraging insect diversity.

We can finish with a summary of the key messages from each chapter:

1   All gardens contribute to wildlife – and they do not need to look wild. Native plants are excellent, but a mix of plants, native and non-native, is a good gardener's compromise and will serve the needs of a variety of species.
2   Gardens, of whatever kind, will have many more species than are ever seen by gardeners (especially invertebrate species).

There will be many hundreds of species present in the garden or just passing through.

3   Gardens with a variety of habitat types are likely to harbour more species. Larger gardens generally have more trees and so gain species particularly associated with those trees.

Because many garden species are mobile, differences in species richness recorded in adjacent gardens with different features are less pronounced than might be expected.

Green roofs generally seem to do little to add to biodiversity, while mini-meadows and tapestry lawns are likely to increase invertebrate abundance and possibly species richness.

4   There are a variety of terms used to describe different categories of plants found in the UK. The fundamental distinction is between plants that have 'always' been in the UK. This means since before the severing of the land bridge with mainland Europe around 8,000 years ago. Such plants are referred to as native – those which have colonized in more recent times are non-native.

5   Why do animals eat some plants or are attracted to some plants but not others? There may be aspects of the plant that either attract or repel. For example, plants can contain a variety of molecules known as secondary metabolites which can inhibit invertebrate feeding. Colour or scent may attract pollinators. Plant breeders can alter the association between plants and insects (or other invertebrates). New varieties with double flowers may be inaccessible to pollinators, but plant varieties may also show more subtle changes, such as flowering time, scent and so on.

6   Gardens are ecosystems with a very large variety of interacting species linked through a food web. Greater biodiversity is generally considered to result in greater ecosystem stability, which is good for the garden.

7   There is wide recognition of the range of ecosystems services provided by nature. On a smaller scale, the same applies in

gardens – where natural processes such as 'pest control' and decomposition reflect the functioning of the garden as an ecosystem. 'Cultural' services provided by gardens include contributing to human well-being. Gardens can also collectively contribute to ecosystem services on an urban scale, such as ameliorating summer temperatures (where trees provide shade) and contributing to infiltration of rainwater.

8 In terms of biodiversity, a garden may not harbour dormice or deer, but it will have birds, which are easy to spot, as well as a wide range of plants. However, the real core of biodiversity in the garden (in terms of species number) will come from the small things. It is the invertebrates and microorganisms that can be astonishingly varied. The microorganisms are very much for specialist study (apart from the larger macrofungi – mushrooms and toadstools). But a number of amateur studies have demonstrated that with patience, time and effort it is possible to record hundreds or even thousands of invertebrate species in a garden. These studies have confirmed that gardens need not be seen as biodiversity deserts.

9 Soils make their contribution to garden biodiversity. Yet soils get some 'rough' treatment in the garden – digging is the most obvious – which potentially disrupts the fabric of the soil ecosystem. Despite this, limited studies on garden soils suggest a fair degree of resilience of the living soil community. At the same time, with only limited studies of the impact of various gardening activities on garden soil, there is much yet to discover. Beyond diversity, the nature of the soil – especially fertility – has a big impact on the options for creating small, flower-rich 'meadows' in gardens. High soil fertility generally decreases plant-species richness.

10 Why use chemicals against 'pests' when you can get a variety of natural enemies to do the job for you? Of course, it's not quite as simple as that. Natural enemies are a mix of invertebrates and microorganisms that will naturally kill your pests. Many of these natural enemies are parasitoid insects. Others

are predatory insects. Species in both groups can be attracted into the garden by a flower-rich mix of plants, and then feed on nectar, pollen and honeydew. There will be a reduction in pest numbers, but this may not be sufficient to avoid some pest damage. But the absence of biocides should be positive for biodiversity in the garden as a whole.

11 Pollinating insects provide a critical ecosystem service in the garden as well as enhancing its biodiversity. The secret of keeping them 'happy' is to have a flower-rich garden. There has been much debate about the value of native versus non-native plant species. The consensus is that a good mix of flower types is fine – the non-natives are useful because they often flower earlier or later in the season. This chapter also dealt with herbivorous insects – and again the biodiversity of these is enhanced by a good mix of flowering species, both native and non-native. Flowers of the plant family Asteraceae/Compositae are the most all-round useful.

12 Rewilding/wilding is a fashionable concept in conservation but in its true sense is only peripherally relevant to gardeners with an average garden. Nonetheless, there is certainly something to be said for having a wilder, less managed part of the garden or lawn. This will provide additional niche opportunities for a range of plants and invertebrates.

So, no magic is required to improve the 'attractiveness' of your garden to a variety of species and so enhance biodiversity. The impact from a biodiversity-friendly individual garden may be small but, taken collectively, UK gardens occupy a large area of land. There is some variation in figures quoted, but according to the UK Office for National Statistics, private gardens in the UK occupy an area of approximately 522,000 hectares/1,289,890 acres of land or 5,220 square kilometres/2,015 square miles. This is more than four and a half times greater than the area of all the UK National Nature Reserves put together. For urban environments taken as a whole, gardens occupy about 30 per cent of the land area. If more gardens had some element of wilding and/or floral enrichment, this could be a big win for biodiversity as well as garden-orientated ecosystem services. The time has come to put away the pesticide sprayer, to hand weed and apply 'organic' methods of 'pest' control only where necessary, make a compost heap and log pile, enhance the organic content of the soil, leave some areas to go wild(er), plant native or other wildlife-friendly trees, enhance the floral richness of the garden beds (native and non-native) and plant those mini-meadows. Easy, really...

Appendix
Survey Methods for Garden Wildlife

This Appendix gives a brief description of commonly used and relatively straightforward survey methods for sampling a range of wildlife species, all of which can be used in a garden setting. They can be applied in a non-quantitative way by gardeners who would just like to know what is around in their garden – rather than a detailed quantitative scientific survey. For some species – butterflies, for instance – it may not be necessary to capture the organism; it will suffice to just spend time in the garden looking for what can be found with a field guide or identification chart in hand. A photograph can help in cases of doubt and many mobile phones can take decent macro photographs – if you can get close enough. However, with many species this will not be sufficient and you will need to capture the organism to allow identification. Some of the invertebrate trapping methods described will result in a dead specimen, which is necessary if fine detail needs to be checked with a lens or microscope.

Bat detectors: Although we have not specifically discussed bats in any chapter, they are an integral part of the garden eco-system – if you are lucky enough to have them. They can be seen flying at dusk during the warmer months but are difficult to identify just by observation. The high-frequency sounds they emit for echolocation can be converted into audible sounds and the frequency captured with a bat detector. You can then check the frequency against a standard list. The most costly types of detector will have a display to show the likely species of bat. Expensive.

Beating tray: A pale-coloured cloth is stretched over a frame and held under a tree or shrub. Insects and other invertebrates are dislodged by 'beating' the branches of the tree or shrub. The 'catch' of invertebrates can be picked off from the cloth and transferred to sample tubes for checking – or they can be sucked up by a pooter (see later entry). This needs to be done swiftly – the insects don't stay around for long!

Butterfly net: A fine and relatively long (50cm/20in) mesh bag, with an open frame and handle, used for capturing butterflies in flight. To avoid the butterfly escaping, the net is folded over after capture, sandwiching the butterfly between the two layers of net. The wings of the live butterfly can then be examined for a species identification before the butterfly is released. Moderately cheap.

Camera trap: These aids to locating shy and night-active mammals have become essential kit. The camera is set up overnight, is triggered by movement and will take a photograph or video using more or less invisible infrared light. Depending on how and where the camera trap is set up, it may photograph mammals ranging in size from mice and voles to badgers and deer – as well as cats! Images are stored on a memory card and can be examined the following day on a computer. Relatively expensive.

Lens (loupe)/microscope: Some species will require detailed examination of specific features that may help distinguish species using a published identification key. For some species a small lens (loupe) will suffice; this is often the case with plants. For others, a low-powered microscope (known as a stereo/dissecting microscope) with a strong light source is essential: for example, with ground beetles, which require examination of the arrangement of hairs (setae) on the exoskeleton. Lens (loupes) vary from fairly cheap to fairly expensive. Microscopes are expensive.

Malaise trap: This large, partially open-sided tent-like structure is made of netting. It has two short end walls, one central wall and a roof, which is higher at one end. When insects fly into the structure they are stopped by a central wall of netting and then fly upwards to the higher apex, where they end up in a collecting jar. This trap is particularly useful for passive trapping of flying insects, especially (but not only) those belonging to the Hymenoptera (bees, wasps and ants) and Diptera (flies). Typically, the catch is collected in 70 per cent alcohol. Expensive.

Moth traps: These traps are set up at night and use a near ultraviolet light designed to attract moths. They can be battery or mains powered. In fact, moth traps will typically capture other species as well, such as caddis flies (Trichoptera), beetles (Coleoptera) and true bugs (Hemiptera). Captured moths (plus other species) collect in the base of the trap and can be removed in the morning. Moths can be stored for a few hours in a fridge if necessary – they are easier to examine if cold and inactive. They are released alive in the evening after checking. Expensive.

Pan traps: These traps are based on the simple principle that flying insects are attracted to particular colours. Wearing a yellow T-shirt in the summer demonstrates this on warm sunny days by attracting many pollen beetles. The traps consist of a pan of some kind, which has been painted on the inside. A good colour is yellow, but green, blue and red can be used to compare attractiveness to different insects. The trap is filled with water (plus a little detergent) so insects caught will be dead. Cheap.

Pitfall traps: Pitfall traps are used to sample epigeal invertebrates – those that wander over the soil surface. At its simplest, a jar or plastic cup is dug into the soil, with the rim just slightly lower than the soil surface. The trap has a few centimetres of water (plus a little detergent) to trap the insects that are caught dead. It is possible to use traps without water, but predatory insects tend to eat the others! Some kind of cover is needed to stop the trap overflowing in case of rain, but with a gap to allow invertebrates to walk underneath into the trap. Cheap.

Plastic pots and tubes: A selection of pots and tubes are needed for retaining specimens (for example, moths from a moth trap) for identification. A small artists' brush and tweezers or forceps can be useful for handling insects. Cheap.

Pooter: This is a small device for sucking up small invertebrates from nets or other surfaces. It consists of a glass or plastic tube/vial/container with two flexible tubes attached through the tube lid or stopper. One tube is sucked by mouth and can be used to draw up insects with the other tube. The key thing is to have a piece of mesh over the tube end of the sucking tube, or you will end up with a mouthful of insects. There are diagrams on the Web showing how a pooter can easily be constructed. Cheap.

Sweep net: This is a framed net with a short handle with a strong cotton bag rather than a mesh. It is not really a net in the conventional sense. The net is used for swishing through grass and shrubs and lower branches of trees to sample insects and other invertebrates from growing vegetation. Both frame and net need to be tough. The live catch can either be observed in the net or else tipped on to a white sheet or tray. Unless transferred into closed containers, the catch quickly walks or flies away. Moderately cheap.

White tray: This can be used for sorting samples of leaf litter and compost or for emptying a sweep net. A variety of invertebrates are likely to be seen against the white background of the tray. If you have sharp eyes, then tiny compost or soil inverte-brates like mites and springtails can also be spotted. These can be 'hoovered' up with a pooter (see above entry). Cheap.

Glossary

Alien plants: Plants that are considered to have established (in the UK or elsewhere) from abroad.

Anecic: Deep-soil-burrowing earthworms, with burrows to a depth of 1 metre/3 feet or more. Found below normal digging or ploughing depth.

Aphid: Commonly known as greenfly. These are sap-sucking insects that belong to the insect order Hemiptera – the true bugs. They are characterized by needle-like mouthparts and specialize in fluid feeding on plant sap (some other Hemiptera species take animal body fluids).

Archaeophytes: Essentially alien plants (introduced from outside the UK) but established since before around 1500. Now generally considered to be 'honorary' natives.

Aromatics: Organic chemicals that are synthesized in plants and are referred to as plant secondary metabolites. Some of these chemicals can be perceived as scents – hence the term aromatic – which may be present in the flowers, fruit or foliage.

Beneficial insects: Insects that perform services that benefit gardeners – such as pollinating flowers or killing pest species.

Biocides: A chemical (normally) that kills various organisms. Covers the terms pesticides, fungicides, molluscicides, herbicides and insecticides.

Biodiversity: A term used rather loosely to refer to the total assemblage of species at a particular location, such as a garden.

Bryophyte: One group of what are referred to as 'lower plants' – in this case liverworts and mosses. Generally small in size, they have no proper roots or a sap transport (vascular) system for water, minerals and sugars.

Companion planting: A system of pairing two (or more) different plant species, one of which is a crop and which benefits in some way from the non-crop species. Hence it is a one-sided relationship. Typically, such planting is with reference to reducing pest attacks on crop plants, although it could be for other reasons which benefit the crop.

Corolla: A flower typically has a variety of structures inserted in discrete whorls. The whorl of petals, which may be separate or fused, are collectively known as the corolla.

Decomposers: A diverse range of invertebrates and micro-organisms, which collectively form the decomposer community and decompose organic matter such as leaf litter.

Ecology: The scientific study of the interactions between living organisms and with their environment.

Ecosystem: In simple terms this is a semi-independent ecological unit such as a woodland, a pond or even a single plant. It can also be defined as a biological community of interacting organisms and their physical environment.

Ecosystem services: The benefits to human beings provided by ecosystems and their associated biodiversity.

Endogeic: With reference to earthworms, these are species that make horizontal burrows more or less within the top 20cm/8in of soil.

Epigeic: With reference to earthworms, these are species that feed within the layer of organic material on the surface of soil.

Food web: The complex web of feeding connections within an ecosystem, starting with plants and organic detritus and ending with the top predators.

Genes: Units of DNA on chromosomes that code for (or contribute towards) specific characteristics of the organism.

Herbivores: Species that feed on plants – from large mammals to tiny invertebrates like aphids.

Herbicide: A chemical agent (usually) used to kill weeds. May be selective, killing only certain kinds of plants, or non-selective, killing all plants.

Lichen: A composite organism in which the main part is fungal tissue, but with algae and bacteria as symbiotic partners. Very resistant to adverse environmental conditions.

Metabolites: In the complex biochemistry of an organism with a myriad of reactions and chemical transformations, metabolites are the individual kinds of molecule that play a role in metabolism.

Native plants: Plant species that survived the last Ice Age in the UK or arrived in the UK after the last Ice Age but before the severing of the land bridge with mainland Europe (around 8,000 years ago) as sea levels rose.

Natural enemies: A term applied to animal or microbial species that happen to predate or kill insect species, including those considered to be 'pests'.

Neophytes: Plants that have arrived in the UK after 1500. They are alien plants but may have been in the UK for many years or arrived more recently. Some have integrated into the UK flora and become naturalized. Most plants cultivated in gardens since 1500 could be classed as neophytes.

Palynology: Primarily a term used in the context of studying pollen, especially in the context of fossil pollen, which contributes to the interpretation of historical vegetation changes at a particular location.

Parasites: Species that live on or in another host organism and exploit its resources, causing harm to the host. This can be a long-term association that typically does not directly kill the host.

Parasitoids: These are insects that lay their eggs on or in a host species (usually an insect). The immature larvae slowly consume the host and eventually kill it.

Pollinators: Species that inadvertently transfer pollen between plant species while they collect pollen as a food source or sip nectar. The main emphasis is on day-active insects such as honeybees, bumblebees or solitary bees. In fact, there are various other pollinating insects, including night-active moths. In other parts of the world birds, bats and other mammals may also act as pollinators.

Rewilding: This and the related term 'wilding' relate to 'returning the land to nature'. But the extent of this will vary in ambition and scale. In most cases the term covers giving nature partial autonomy, but with varying degrees of continuing management input. The balance between the two will vary from project to project. And the scale will vary – perhaps converting just a small patch of lawn in the case of a small garden.

Saproxylic: Organisms associated with dead wood, which can serve as a habitat, breeding site and food source.

Species richness: The total species count at a particular location. The bigger the number, the more diverse the community.

Tapestry lawn: Planting a mix of herbaceous plants instead of grass to create a lawn. These need to be plants that tolerate trampling and lawn mowing but will still flower.

Taxon: A related group of species, which can be at different levels of a classification. A taxon might, for example, be insects (Class: Insecta); or it might be a smaller group within the insects (Order: Hymenoptera – ants, bees and wasps); or a smaller group still (Genus: *Bombus* – bumblebees).

Taxonomy: The science of classifying organisms into groups based on physical form and/or DNA analysis.

Weed: A plant growing in the wrong place.

Wildlife: A collective term that typically refers to the undomesticated animals (fauna) of an area, although it can rightfully also apply to the plants (flora).

Xerophyte: A plant adapted to living in water-limited environments, such as sand or shingle (the plant is described as xerophytic or xerophilic).

References

Alexander, K. et al. (2006) 'The value of different tree and shrub species to wildlife', *British Wildlife*, Volume 18, pp. 18 –28.

Angold, P. G. et al. (2006) 'Biodiversity in urban habitat patches', *Science of the Total Environment*, Volume 360(1–3), pp. 196–204.

Anthony, M. A. et al. (2023) 'Enumerating soil biodiversity', *Proceedings of the National Academy of Sciences*, Volume 120(33), e2304663120.

Aronson, M. F. et al. (2017) 'Biodiversity in the city: key challenges for urban green space management', *Frontiers in Ecology and the Environment*, Volume 15(4), pp. 189–196.

Asplund, J. and Wardle, D. A. (2017) 'How lichens impact on terrestrial community and ecosystem properties', *Biological Reviews of the Cambridge Philosophical Society*, Volume 92(3), pp. 1720–1738.

Balfour, E. B. (1943) *The Living Soil.* London: Faber and Faber.

Bornkamm, R. et al. (1982) *Urban Ecology: The Second European Ecological Symposium.* Berlin, Oxford: Blackwell Scientific Publications.

Buglife (2023) 'Main groups of insects'. Available at: www.buglife.org.uk/bugs/bug-identification-tips/main-groups-of-insects (accessed: 25.01.23).

Burks, J. M. and Philpott, S. M. (2017) 'Local and landscape drivers of parasitoid abundance, richness, and composition in urban gardens', *Environmental Entomology*, Volume 46(2), pp. 201–209.

Byrne, L. B. and Bruns, M. A. (2004) 'The effects of lawn management on soil microarthropods', *Journal of Agricultural and Urban Entomology*, Volume 21(3), pp. 150–156.

Campbell, A. J. et al. (2012) 'Realizing multiple ecosystem services based on the response of three beneficial insect groups to floral traits and trait diversity', *Basic and Applied Ecology*, Volume 13(4), pp. 363–370.

Camps-Calvet, M. et al. (2016) 'Ecosystem services provided by urban gardens in Barcelona, Spain: Insights for policy and planning', *Environmental Science and Policy*, Volume 62, pp. 14–23.

Chinery, M. (1977) *The Natural History of the Garden*. London: Collins.

Corbet, S. A. et al. (2001) 'Native or exotic? Double or single? Evaluating plants for pollinator-friendly gardens', *Annals of Botany*, Volume 87(2), pp. 228–232.

Darwin, C. (1872, 6th edn.), *On the Origin of Species*. London: John Murray.

Davis, B. N. K. (1991) *Insects on Nettles*. Slough: Richmond Publishing.

De Bell, S. et al. (2020) 'Spending time in the garden is positively associated with health and wellbeing: Results from a national survey in England', *Landscape and Urban Planning*, Volume 200, 103836.

Doberski, J. (2022) *The Science of Compost: Life, Death and Decay in the Garden*. London: Pimpernel Press Ltd.

Egerer, M. H. et al. (2018) 'Cityscape quality and resource manipulation affect natural enemy biodiversity in and fidelity to urban agroecosystems', *Landscape Ecology*, Volume 33(6), pp. 985–998.

Egerer, M. and Philpott, S. M. (2022) '"Tidy" and "messy" management alters natural enemy communities and pest control in urban agroecosystems', *PLoS ONE*, Volume 17(9), e0274122.

Finch, S. et al. (2003) 'Companion planting – do aromatic plants disrupt host-plant finding by the cabbage root-fly and the onion-fly more effectively than non-aromatic plants?: Companion planting and pest insects', *Entomologia Experimentalis et Applicata*, Volume 109(3), pp. 183–195.

Fuller, R. A. et al. (2007) 'Psychological benefits of greenspace increase with biodiversity', *Biology Letters*, Volume 3(4), pp. 390–394.

Fuller, R. and Gilroy, J. (2021) 'Rewilding and intervention: Complementary philosophies for nature conservation in Britain', *British Wildlife*, Volume 32, pp. 258–267.

Garbuzov, M. et al. (2017) 'Most ornamental plants on sale in garden centres are unattractive to flower-visiting insects', *PeerJ*, Volume 5, e3066.

Gardiner M. M. et al. (2014) 'Vacant land conversion to community gardens: Influences on generalist arthropod predators and biocontrol services in urban greenspaces', *Urban Ecosystems*, Volume 17(1), pp. 101–122.

Gaston, K. J. et al. (2004) 'Gardens and wildlife – the BUGS project', *British Wildlife*, Volume 16, pp. 1–9.

Gaston, K. J. et al. (2007) 'Improving the contribution of urban gardens for wildlife: Some guiding propositions', *British Wildlife*, Volume 18, pp. 171–177.

Griffiths-Lee, J. et al. (2020) 'Companion planting to attract pollinators increases the yield and quality of strawberry fruit in gardens and allotments', *Ecological Entomology*, Volume 45(5), pp. 1025–1034.

Griffiths-Lee, J. et al. (2022) 'Sown mini-meadows increase pollinator diversity in gardens', *Journal of Insect Conservation*, Volume 26(2), pp. 299–314.

Hicks, D. M. et al. (2016) 'Food for pollinators: Quantifying the nectar and pollen resources of urban flower meadows', *PLoS ONE*, Volume 11(6), e0158117.

Hill, L. (2022a) 'The other pandemic: Ten years of ash dieback', *British Wildlife*, Volume 34, pp. 113–122.

Hill, M. (ed.) (2022b) *The Nature of Cambridge*. Newbury: Pisces Publications.

Hodge, T. (2022) 'Tim Hodge's garden, Horsey, East Norfolk', *Pan-Species Listing*. Available at: https://psl.brc.ac.uk/content/tim-hodges-garden-horsey-east-norfolk (accessed: 17.08.23).

Knapp, S. et al. (2019) 'Biodiversity impact of green roofs and constructed wetlands as progressive eco-technologies in urban areas', *Sustainability*, Volume 11(20), 5846.

Livesley, S. J. et al. (2016) 'The urban forest and ecosystem services: Impacts on urban water, heat and pollution cycles at the tree, street, and city scale', *Journal of Environmental Quality*, Volume 45(1), pp. 119–124.

Lowenstein, D. M. and Minor, E. S. (2018) 'Herbivores and natural enemies of brassica crops in urban agriculture', *Urban Ecosystems*, Volume 21(3), pp. 519–529.

McIntyre, N. E. et al. (2001) 'Ground arthropod community structure in a heterogeneous urban environment', *Landscape and Urban Planning*, Volume 52(4), pp. 257–274.

MacLeod, A. et al. (2004) 'Beetle banks as refuges for beneficial arthropods in farmland: Long-term changes in predator communities and habitat', *Agricultural and Forest Entomology*, Volume 6(2), pp. 147–154.

Majewska, A. A. and Altizer, S. (2020) 'Planting gardens to support insect pollinators', *Conservation Biology*, Volume 34(1), pp. 15–25.

Mancini, F. et al. (2023) 'Invertebrate biodiversity continues to decline in cropland', *Proceedings of the Royal Society, B,* Volume 290, 20230897.

Matteson, K. C. and Langellotto, G. A. (2010) 'Determinates of inner city butterfly and bee species richness', *Urban Ecosystems*, Volume 13(3), pp. 333–347.

Metal, P. (2017) *Insectinside: Life in the Bushes of a Small Peckham Park*. London: Independent Publishing Network.

Millenium Ecosystem Assessment (2005) *Ecosystems and Human Well-Being: Synthesis*. Washington, DC: Island Press.

Norton, B. A. et al. (2014) 'The effect of urban ground cover on arthropods: An experiment', *Urban Ecosystems*, Volume 17(1), pp. 77–99.

Ollerton, J. et al. (2011) 'How many flowering plants are pollinated by animals', *Oikos,* Volume 120(3), pp. 321–326.

Orros, M. E. et al. (2015) 'Supplementary feeding of wild birds indirectly affects ground beetle populations in suburban gardens', *Urban Ecosystems,* Volume 18(2), pp. 465–475.

Owen, J. (2010) *Wildlife of a Garden: A Thirty-Year Study*. London: Royal Horticultural Society.

Owen, J. (2015) Jennifer Owen's Leicester garden', *Pan Species Listing*. Available at: https://psl.brc.ac.uk/content/jennifer-owens-leicester-garden (accessed: 17.08.23).

Perrone, A. (2023) 'Cacti replacing snow on Swiss mountainside due to global heating', *The Guardian*. Available at: https://www.theguardian.com/environment/2023/feb/10/cacti-replacing-snow-on-swiss-mountainsides-due-to-global-heating (accessed: 17.08.23).

Philpott, S. M. and Bichier, P. (2017) 'Local and landscape drivers of predation services in urban gardens', *Ecological Applications*, Volume 27(3), pp. 966–976.

Plummer, K. E. et al. (2019) 'The composition of British bird communities is associated with long-term garden bird feeding', *Nature Communications*, Volume 10, 2088.

Pocock, M. J. O. et al. (2023) 'The benefits of citizen science and nature-noticing activities for well-being, nature connectedness and pro-nature conservation behaviours', *People and Nature*, Volume 5(2), pp. 591–606.

RHS (2021) *The RHS Sustainability Strategy: Net Positive for Nature and People by 2030*. London: Royal Horticultural Society.

RHS (2022a) *Plant Finder*. London: Royal Horticultural Society Media.

RHS (2022b) 'Why the p-word has to go', *The Garden*, June 2022, 9.

Rocha, E. A. et al. (2018) 'Influence of urbanisation and plants on the diversity and abundance of aphids and their ladybird and hoverfly predators in domestic gardens', *European Journal of Entomology*, Volume 115(1), pp. 140–149.

Rollings, R. and Goulson, D. (2019) 'Quantifying the attractiveness of garden flowers for pollinators', *Journal of Insect Conservation*, Volume 23(5–6), pp. 803–817.

Salisbury, A. et al. (2015) 'Enhancing gardens as habitats for flower-visiting aerial insects (pollinators): Should we plant native or exotic species?', *Journal of Applied Ecology*, Volume 52(5), pp. 1156–1164.

Salisbury, A. et al. (2017) 'Enhancing gardens as habitats for plant-associated invertebrates: Should we plant native or exotic species?', *Biodiversity and Conservation*, Volume 26(11), pp. 2657–2673.

Salisbury, A. et al. (2020) 'Enhancing gardens as habitats for soil-surface-active invertebrates: Should we plant native or exotic species?', *Biodiversity and Conservation*, Volume 29(1), pp. 129–151.

Sánchez-Bayo, F. and Wyckhuys, K. A. G. (2019) 'Worldwide decline of the entomofauna: A review of its drivers', *Biological Conservation*, Volume 232, pp. 8–27.

Schulte to Bühne, H. et al. (2022) 'The policy consequences of defining rewilding', *Ambio*, Volume 51(1), pp. 93–102.

Seitz, N. et al. (2020) 'Are native and non-native pollinator friendly plants equally valuable for native wild bee communities?', *Ecology and Evolution*, Volume 10(23), pp. 12838–12850.

Silvertown, J. et al. (2006) 'The Park Grass Experiment 1856–2006: Its contribution to ecology', *Journal of Ecology*, Volume 94(4), pp. 801–814.

Siviter, H. et al. (2021) 'Agrochemicals interact synergistically to increase bee mortality', *Nature*, Volume 596(7892), pp. 389–392.

Smith, L. S. (2019) *Tapestry Lawns: Freed from Grass and Full of Flowers*. Boca Raton: CRC Press.

Smith, L. S. et al. (2015) 'Adding ecological value to the urban lawnscape: Insect abundance and diversity in grass-free lawns', *Biodiversity and Conservation*, Volume 24(1), pp. 47–62.

Smith, R. M. et al. (2005) 'Urban domestic gardens (V): Relationships between landcover composition, housing and landscape', *Landscape Ecology*, Volume 20(2), pp. 235–253.

Smith, R. M. et al. (2006a) 'Urban domestic gardens (VI): Environmental correlates of invertebrate species richness', *Biodiversity and Conservation*, Volume 15(8), pp. 2415–2438.

Smith, R. M. et al. (2006b) 'Urban domestic gardens (VIII): Environmental correlates of invertebrate abundance', *Biodiversity and Conservation*, Volume 15(8), pp. 2515–2545.

Smith, R. M. et al. (2006c) 'Urban domestic gardens (IX): Composition and richness of the vascular plant flora, and implications for native biodiversity', *Biological Conservation*, Volume 129(3), pp. 312–322.

Smith, R. M. et al. (2010) 'Urban domestic gardens (XIII): Composition of the bryophyte and lichen floras, and determinants of species richness', *Biological Conservation*, Volume 143(4), pp. 873–882.

Tallamy, D. W. et al. (2010) 'Can alien plants support generalist insect herbivores?', *Biological Invasions*, Volume 12(7), pp. 2285–2292.

Tanaka, S. and Ohsaki, N. (2006) 'Behavioural manipulation of host caterpillars by the primary parasitoid wasp *Cotesia glomerata* (L.) to construct defensive webs against hyperparasitism', *Ecological Research*, Volume 21(4), pp. 570–577.

Tassin de Montaigu, C. and Goulson, D. (2023) 'Habitat quality, urbanisation & pesticides influence bird abundance and richness in gardens', *Science of the Total Environment*, Volume 870, 161916.

Tew, N. E. et al. (2022) 'Turnover in floral composition explains species diversity and temporal stability in the nectar supply of urban residential gardens', *Journal of Applied Ecology*, Volume 59(3), pp. 801–811.

Thomas, A. (2017) *Gardening for Wildlife*. London: Bloomsbury Publishing.

Thompson, K. (No Year) 'Foodwebs and feeding roles', *Wildlife Gardening Forum*. Available at: wlgf.org/food_webs.html (accessed: 23.02.23).

Thompson, K. et al. (2003) 'Urban domestic gardens (I): Putting small-scale plant diversity in context', *Journal of Vegetation Science*, Volume 14(1), pp. 71–78.

Thompson, K. et al. (2004) 'Urban domestic gardens (III): Composition and diversity of lawn floras', *Journal of Vegetation Science*, Volume 15(3): pp. 373–378.

Thompson, K. et al. (2005) 'Urban domestic gardens (VII): A preliminary survey of soil seed banks', *Seed Science Research*, Volume 15(2), pp. 133–141.

Thuring, C. and Grant, G. (2016) 'The biodiversity of temperate extensive green roofs – a review of research and practice', *Israel Journal of Ecology & Evolution*, Volume 62(1–2), pp. 44–57.

Torppa, K. A. and Taylor, A. R. (2022) 'Alternative combinations of tillage practices and crop rotations can foster earthworm density and bioturbation', *Applied Soil Ecology*, Volume 175, 104460.

Tratalos, J. et al. (2007) 'Urban form, biodiversity potential and ecosystem services', *Landscape and Urban Planning*, Volume 83(4), pp. 308–317.

Tree, I. (2018) *Wilding: The Return of Nature to a British Farm*. London: Picador.

Tresch, S. et al. (2018) 'A gardener's influence on urban soil quality', *Frontiers in Environmental Science*, Volume 6, 25.

Tschumi, M. et al. (2015) 'High effectiveness of tailored flower strips in reducing pest and crop plant damage', *Proceedings of the Royal Society, B*, Volume 282(1814), pp. 1–8.

UK National Ecosystem Assessment (2011) *The UK National Ecosystem Assessment: Synthesis of the Key Findings*. Cambridge: UNEP-WCMC.

Walliser, J. (2020) *Plant Partners: Science-Based Companion Planting Strategies for the Vegetable Garden*. North Adams, MA: Storey Publishing.

Williams, N. S. G. et al. (2014) 'Do green roofs help urban biodiversity conservation?', *Journal of Applied Ecology*, Volume 51(6), pp. 1643–1649.

Wright, S. I. et al. (2013) 'Evolutionary consequences of self-fertilization in plants', *Proceedings of the Royal Society, B*, Volume 280, 20130133.

WWF (2018) *Living Planet Report – 2018: Aiming Higher*, ed. M. Grooten and R. E. A. Almond. Gland, Switzerland: WWF.

Index

climax vegetation 93

common poppy 32, 87, *88*

companion planting 26–7

compost 67

  heap 2, 3, 16, 100, 106

  invertebrates 64

conservation 13, 24, 52, 53, 54, 60, 94

Convention on Biodiversity 22

creeping cinquefoil 29, 31, 32

cropland areas 6

cross-pollination 38, 82

cultural services *45*, 46 *47*, *50*, 104

dark-edged bee-fly 42

decomposer *41*, 65, 67, 69

deflected succession 93

density of addresses *49*

double flowers 36, 37, *37*, 38, 103

drought 23, 24

earthworm species 65, *66*

ecological succession 93

ecosystem services 6, 44–51, *45*, *47*,
  50, 82, 103, 104, 106

elaiosome 69

emerald ash borer beetle 61

endogeic worms 65, *66*, 67

epigeic worms 65, *66*

exotic plants 2, 33, 34, 68–9, 82–4,
  89–91, 96

flower

  beds 12, 16, 18, 31, 49

  breeding 36

  colour 25, 36

  double 36, 37, *37*, 38, 103

  meadow 70, 71, 85, 93, 95, 104

  morphology 26

  native *17*, 100, 105

  non-native 81, 105

  structure 37, 38, 85, *86*

  wild 28, 30, 95, 101

food web 24, 40, *41*, 42, 65, 68, 103

garden

  as an ecosystem 3, 4, 5, 7, 39–43,
    41, 53, 72

  biodiverse 1, 22, 25, 105

  boundaries 18

  city 5

  complexity 3, 4

  design 3, 7, 39, 41

  ecology 4, 10, 13, 14, 35, 46, 72,
    83

  flora 30–34

  green roofs 22–4

  insect diversity *58*

  insect-friendly 38, 82

  management 7, 39, 41, 72, 76,
    95, 99

  neglected 2

  plants 2, 8, 30, 35, 36, *41*, 69, 82,
    84

  rock 15, 33

  size 16, 18, 49, 55, 87

  studies 9, 40, 54, 55

  trees *50*, 51

  wildlife 1–7, 9, 12, 27, 81, 91,
    107–11

  zones 16

Acknowledgements

Gardens were, for a long time, a Cinderella subject in the science of ecology. They were thought of as too artificial and too impoverished of nature, hence primarily of interest to horticulturalists. One study of animal and plant diversity in a suburban garden in the UK city of Leicester contributed markedly to a change of perspective. The author of this thirty-year study, Jennifer Owen, made the study of garden biodiversity and ecology scientifically interesting. This book is based on her work and many others who have followed in her footsteps, studying gardens and urban environments. Among these is Tim Hodge, who is thanked for providing me with updated species data for his garden in Norfolk (UK). The University of University of Waikato (NZ) Science Learning Hub is thanked for permission to use their nice diagram of a parasitoid life cycle, as well as other authors whose figures were adapted for the book.

The first draft of my book was read and edited by my wife, Christine, followed by further editing and design work by Anna Sanderson, Monica Hope and Sarah Pyke at Pimpernel Press, with illustrations by Thomas Bohm. Their collective contributions and efforts have greatly enhanced the book.